the hidden connections

DOUBLEDAY

New York London Toronto Sydney Auckland

the hidden connections

INTEGRATING THE BIOLOGICAL, COGNITIVE, AND SOCIAL DIMENSIONS OF LIFE INTO A SCIENCE OF SUSTAINABILITY

Fritjof Capra

PUBLISHED BY DOUBLEDAY
A division of Random House, Inc.
1540 Broadway, New York, New York 10036

DOUBLEDAY is a trademark of Doubleday, a division of Random House, Inc.

Book design by Nicola Ferguson

Library of Congress Cataloging-in-Publication Data
Capra, Fritjof.
The hidden connections: integrating the biological, cognitive, and social dimensions of life into a science of sustainability / Fritjof Capra.—1st ed.
p. cm.
Includes bibliographical references and index.
1. New Age movement. I. Title.
BP605.N48 C365 2002
191—dc21 2002023352

ISBN 0-385-49471-8

PRINTED IN THE UNITED STATES OF AMERICA

FIRST EDITION

September 2002

1 3 5 7 9 10 8 6 4 2

To Elizabeth and Juliette

Education is the ability to perceive
the hidden connections between phenomena.

—Václav Havel

| contents |

| acknowledgments |

For the past twenty-five years, I have practiced a style of research that relies heavily on dialogues and discussions with individuals and small groups of friends and colleagues. Most of my insights and ideas originated and were further refined in those intellectual encounters. The ideas presented in this book are no exception.

I am especially grateful

- to Pier Luigi Luisi for many stimulating discussions about the nature and origin of life, and for his warm hospitality at the Cortona Summer School in August 1998, and at ETH in Zurich in January 2001;
- to Brian Goodwin and Richard Strohman for challenging discussions about complexity theory and cellular biology;
- to Lynn Margulis for enlightening conversations about microbiology, and for introducing me to the work of Harold Morowitz;
- to Francisco Varela, Gerald Edelman, and Rafael Nuñez for enriching discussions about the nature of consciousness;
- to George Lakoff for introducing me to cognitive linguistics and for many clarifying conversations;

➣ to Roger Fouts for illuminating correspondence about the evolutionary origins of language and consciousness;

➣ to Mark Swilling for challenging discussions about the similarities and differences between the natural and social sciences, and for introducing me to the work of Manuel Castells;

➣ to Manuel Castells for his encouragement and support, and for a series of stimulating systematic discussions of fundamental concepts in social theory, of technology and culture and of the complexities of globalization;

➣ to William Medd and Otto Scharmer for clarifying conversations about social science;

➣ to Margaret Wheatley and Myron Kellner-Rogers for inspiring dialogues over several years about complexity and self-organization in living systems and human organizations;

➣ to Oscar Motomura and his colleagues at AMANA-KEY for continually challenging me to apply abstract ideas to professional education, and for their warm hospitality in São Paulo, Brazil;

➣ to Angelika Siegmund, Morten Flatau, Patricia Shaw, Peter Senge, Etienne Wenger, Manuel Manga, Ralph Stacey and the SOLAR group at Nene Northampton College for numerous stimulating discussions of management theory and practice;

➣ to Mae-Wan Ho, Brian Goodwin, Richard Strohman, and David Suzuki for illuminating discussions of genetics and genetic engineering;

➣ to Steve Duenes for a helpful conversation about the literature on metabolic networks;

➣ to Miguel Altieri and Janet Brown for helping me to understand the theory and practice of agroecology and organic farming;

➣ to Vandana Shiva for numerous inspiring conversations about science, philosophy, ecology, community and the Southern perspective on globalization;

➣ to Hazel Henderson, Jerry Mander, Douglas Tompkins,

and Debi Barker for challenging dialogues about technology, sustainability, and the global economy;

➣ to David Orr, Paul Hawken, and Amory Lovins for many informative conversations about ecodesign;

➣ to Gunter Pauli for extended stimulating dialogues on three continents about the ecological clustering of industries;

➣ to Janine Benyus for a long and inspiring discussion of nature's "technological miracles";

➣ to Richard Register for many discussions about how to apply ecodesign principles to urban planning;

➣ to Wolfgang Sachs and Ernst-Ulrich von Weizsäcker for informative conversations about Green politics;

➣ and to Vera van Aaken for introducing me to a feminist perspective on excessive material consumption.

During the last few years, while I was working on this book, I was fortunate to attend several international symposia where many of the issues I was exploring were discussed by authorities in various fields. I am deeply grateful to Václav Havel, president of the Czech Republic, and to Oldrich Cerny, executive director of the Forum 2000 Foundation, for their generous hospitality at the annual Forum 2000 symposia in Prague in the years 1997, 1999 and 2000.

I am indebted to Ivan Havel, director of the Center for Theoretical Study in Prague, for the opportunity to participate in a symposium on science and teleology at Charles University in March 1998.

I am very grateful to the Piero Manzù International Research Center for inviting me to participate in a symposium on the nature of consciousness in Rimini, Italy, in October 1999.

I am indebted to Helmut Milz and Michael Lerner for giving me the opportunity to discuss recent psychosomatic research with leading experts in the field during a two-day symposium at the Commonwealth Center in Bolinas, California, in January 2000.

I am grateful to the International Forum on Globalization for inviting me to participate in two of their intensive and highly informative teach-ins on globalization in San Francisco (April 1997) and New York (February 2001).

While I was working on this book, I had the valuable opportunity to present tentative ideas to international audiences during two courses at Schumacher College in England during the summers of 1998 and 2000. I am deeply indebted to Satish Kumar and the Schumacher College community for extending their warm hospitality to me and my family, as they have done so often in the past, and to my students in these two courses for countless critical questions and helpful suggestions.

In the course of my work at the Center for Ecoliteracy in Berkeley, I have had many opportunities to discuss new ideas about education for sustainable living with a network of outstanding educators, which has helped me greatly in refining my conceptual framework. I am very grateful to Peter Buckley, Gay Hoagland, and especially to Zenobia Barlow for giving me this opportunity.

I wish to thank my literary agent, John Brockman, for his encouragement, and for helping me to formulate the initial outline of the book.

I am deeply grateful to my brother, Bernt Capra, for reading the entire manuscript and for his enthusiastic support and valuable advice on numerous occasions. I am also very grateful to Ernest Callenbach and Manuel Castells for reading the manuscript and for many critical comments.

I am indebted to my assistant, Trena Cleland, for her superb editing of the manuscript, and for keeping my home office running smoothly while I was fully concentrating on my writing.

I am grateful to my editor Roger Scholl at Doubleday for his advice and support, and to Sarah Rainone for seeing the manuscript through the publishing process.

Last but not least, I wish to express my deep gratitude to my wife, Elizabeth, and my daughter, Juliette, for their patience and understanding during many months of strenuous work.

| preface |

In this book I propose to extend the new understanding of life that has emerged from complexity theory to the social domain. To do so, I present a conceptual framework that integrates life's biological, cognitive and social dimensions. My aim is not only to offer a unified view of life, mind and society, but also to develop a coherent, systemic approach to some of the critical issues of our time.

The book is divided into two parts. In Part One, I present the new theoretical framework in three chapters, which respectively deal with the nature of life, the nature of mind and consciousness and the nature of social reality. Readers who are more interested in the practical applications of this framework should turn to Part Two (Chapters 4–7) right away. These chapters can be read independently, but they are cross-referenced to the relevant theoretical sections for those who wish to go into further depth.

In Chapter 4, I apply the social theory developed in the preceding chapter to the management of human organizations, focusing in particular on the question: to what extent a human organization can be considered a living system.

In Chapter 5, I shift my focus to the world at large to deal with one of the most urgent and most controversial issues of our time—the challenges and dangers of economic globalization under the rules of the

World Trade Organization (WTO) and other institutions of global capitalism.

Chapter 6 is dedicated to a systemic analysis of the scientific and ethical problems of biotechnology (genetic engineering, cloning, genetically modified foods etc.), with special emphasis on the recent conceptual revolution in genetics triggered by the discoveries of the Human Genome Project.

In Chapter 7, I discuss the state of the world at the beginning of our new century. After reviewing some of the major environmental and social problems and their connections with our economic systems, I describe the growing worldwide "Seattle Coalition" of nongovernmental organizations (NGOs) and its plans for reshaping globalization according to different values. The final part of the chapter reviews the recent dramatic rise of ecological design practices and discusses their implications for the transition to a sustainable future.

This represents a continuation and evolution of my previous work. Since the early 1970s, my research and writing have focused on a central theme: the fundamental change of worldview that is occurring in science and in society, the unfolding of a new vision of reality and the social implications of this cultural transformation.

In my first book, *The Tao of Physics* (1975), I discussed the philosophical implications of the dramatic changes of concepts and ideas that occurred in physics, my original field of research, during the first three decades of the twentieth century, which are still being elaborated in our current theories of matter.

My second book, *The Turning Point* (1982), showed how the revolution in modern physics foreshadowed a similar revolution in many other sciences and a corresponding transformation of worldviews and values in society. In particular, I explored paradigm shifts in biology, medicine, psychology and economics. In doing so, I came to realize that these disciplines all deal with life in one way or another—with living biological and social systems—and that the "new physics" was therefore inappropriate as a paradigm and source of metaphors in these fields. The physics paradigm had to be replaced by a broader conceptual framework, a vision of reality in which life was at the very center.

This was a profound change of perception for me, which took place gradually and as a result of many influences. In 1988, I published a personal account of this intellectual journey, titled *Uncommon Wisdom: Conversations with Remarkable People.*

At the beginning of the 1980s, when I wrote *The Turning Point*, the new vision of reality that would eventually replace the mechanistic Cartesian worldview in various disciplines was by no means well articulated. I called its scientific formulation "the systems view of life," referring to the intellectual tradition of systems thinking, and I also argued that the philosophical school of deep ecology, which does not separate humans from nature and recognizes the intrinsic values of all living beings, could provide an ideal philosophical, and even spiritual, context for the new scientific paradigm. Today, twenty years later, I still hold this view.

During subsequent years, I explored the implications of deep ecology and the systems view of life with the help of friends and colleagues in various fields and published the results of our explorations in several books. *Green Politics* (coauthored with Charlene Spretnak, 1984) analyzes the rise of the Green Party in Germany; *Belonging to the Universe* (coauthored with David Steindl-Rast and Thomas Matus, 1991) explores parallels between the new thinking in science and Christian theology; *EcoManagement* (coauthored with Ernest Callenbach, Lenore Goldman, Rüdiger Lutz and Sandra Marburg, 1993) proposes a conceptual and practical framework for ecologically conscious management; and *Steering Business Toward Sustainability* (coedited with Gunter Pauli, 1995) is a collection of essays by business executives, economists, ecologists and others who outline practical approaches to meeting the challenge of ecological sustainability. Throughout these explorations my focus was, and still is, on the processes and patterns of organization of living systems—on the "hidden connections between phenomena."[1]

The systems view of life, as outlined in *The Turning Point*, was not a coherent theory of living systems but rather a new way of thinking about life, including new perceptions, a new language and new concepts. It was a conceptual development at the forefront of science, pioneered by researchers in many fields, that created an intellectual climate in which significant advances would be made in the years to follow.

Since then, scientists and mathematicians have taken a giant step toward the formulation of a theory of living systems by developing a new mathematical theory—a body of mathematical concepts and techniques—to describe and analyze the complexity of living systems. This has often been called "complexity theory" or "the science of complexity" in popular writing. Scientists and mathematicians prefer to call it, more prosaically, "nonlinear dynamics."

In science, until recently, we were taught to avoid nonlinear equations, because they were almost impossible to solve. In the 1970s, however, scientists for the first time had powerful high-speed computers that helped them tackle and solve these equations. In doing so, they developed a number of novel concepts and techniques that gradually converged into a coherent mathematical framework.

During the 1970s and 1980s, the interest in nonlinear phenomena generated a whole series of powerful theories that have dramatically increased our understanding of many key characteristics of life. In my most recent book, *The Web of Life* (1996), I summarized the mathematics of complexity and presented a synthesis of contemporary nonlinear theories of living systems that can be seen as an outline of an emerging new scientific understanding of life.

Deep ecology, too, was further developed and refined during the 1980s, and there have been numerous articles and books about related disciplines, such as eco-feminism, eco-psychology, eco-ethics, social ecology and transpersonal ecology. Accordingly, I presented an updated review of deep ecology and its relationships to these philosophical schools in the first chapter of *The Web of Life.*

The new scientific understanding of life, based on the concepts of nonlinear dynamics, represents a conceptual watershed. For the first time, we now have an effective language to describe and analyze complex systems. Concepts like attractors, phase portraits, bifurcation diagrams and fractals did not exist before the development of nonlinear dynamics. Today, these concepts allow us to ask novel questions, and they have led to important insights in many fields.

My extension of the systems approach to the social domain explicitly includes the material world. This is unusual, because traditionally social scientists have not been very interested in the world of matter. Our academic disciplines have been organized in such a way that the natural sciences deal with material structures while the social sciences deal with social structures, which are understood to be, essentially, rules of behavior. In the future, this strict division will no longer be possible, because the key challenge of this new century—for social scientists, natural scientists and everyone else—will be to build ecologically sustainable communities, designed in such a way that their technologies and social institutions—their material and social structures—do not interfere with nature's inherent ability to sustain life.

The design principles of our future social institutions must be consistent with the principles of organization that nature has evolved to sustain the web of life. A unified conceptual framework for the understanding of material and social structures will be essential for this task. The purpose of this book is to provide a first sketch of such a framework.

Berkeley, August 2002
Fritjof Capra

| part one |

LIFE, MIND, AND SOCIETY

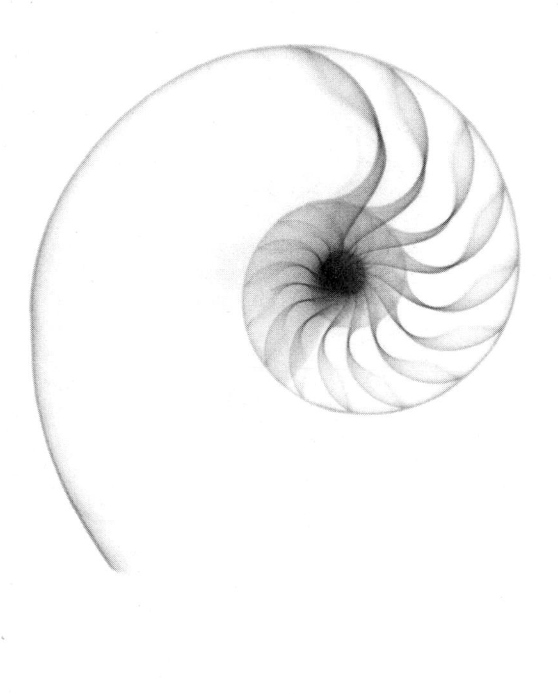

| one |

THE NATURE OF LIFE

Before introducing the new unified framework for the understanding of biological and social phenomena, I would like to revisit the age-old question "What is life?" and look at it with fresh eyes.[1] I should emphasize right from the start that I will not address this question in its full human depth, but will approach it from a strictly scientific perspective; and even then, my focus will at first be narrowed down to life as a biological phenomenon. Within this restricted framework, the question may be rephrased as: "What are the defining characteristics of living systems?"

Social scientists might prefer to proceed in the opposite order—first identifying the defining characteristics of social reality, and then extending into the biological domain and integrating it with corresponding concepts in the natural sciences. This would no doubt be possible, but having been trained in the natural sciences and having previously developed a synthesis of the new conception of life in these disciplines, it is natural for me to begin there.

I could also argue that, after all, social reality evolved out of the biological world between two and four million years ago, when a species of "Southern apes" (*Australopithecus afarensis*) stood up and began to walk on two legs. At that time, the early hominids developed complex brains, toolmaking skills and language, while the helplessness of their prematurely born infants led to the formation of the supportive families and communities that became the foundation of human social life.[2] Hence, it makes sense to ground the understanding of social phenomena in a unified conception of the evolution of life and consciousness.

Focus on Cells

When we look at the enormous variety of living organisms—animals, plants, people, microorganisms—we immediately make an important discovery: all biological life consists of cells. Without cells, there is no life on this Earth. This may not always have been so—and I shall come back to this question[3]—but today we can say confidently that all life involves cells.

This discovery allows us to adopt a strategy that is typical of the scientific method. To identify the defining characteristics of life, we look for and then study the simplest system that displays these characteristics. This reductionist strategy has proved very effective in science—provided that one does not fall into the trap of thinking that complex entities are nothing but the sum of their simpler parts.

Since we know that all living organisms are either single cells or multicellular, we know that the simplest living system is the cell.[4] More precisely, it is a bacterial cell. We know today that all higher forms of life have evolved from bacterial cells. The simplest of these belong to a family of tiny spherical bacteria known as mycoplasm, with diameters less than a thousandth of a millimeter and genomes consisting of a single closed loop of double-stranded DNA.[5] Yet even in these minimal cells, a complex network of metabolic processes* is ceaselessly at work, transporting nutrients in and waste out of the cell, and continually using food molecules to build proteins and other cell components.

*Metabolism, from the Greek *metabole* ("change"), is the sum of biochemical processes involved in life.

Although mycoplasm are minimal cells in terms of their internal simplicity, they can only survive in a precise and rather complex chemical environment. As biologist Harold Morowitz points out, this means that we need to distinguish between two kinds of cellular simplicity.[6] Internal simplicity means that the biochemistry of the organism's internal environment is simple, while ecological simplicity means that the organism makes few chemical demands on its external environment.

From the ecological point of view, the simplest bacteria are the cyanobacteria, the ancestors of blue-green algae, which are also among the oldest bacteria, their chemical traces being present in the earliest fossils. Some of these blue-green bacteria are able to build up their organic compounds entirely from carbon dioxide, water, nitrogen and pure minerals. Interestingly, their great ecological simplicity seems to require a certain amount of internal biochemical complexity.

The Ecological Perspective

The relationship between internal and ecological simplicity is still poorly understood, partly because most biologists are not used to the ecological perspective. As Morowitz explains:

> Sustained life is a property of an ecological system rather than a single organism or species. Traditional biology has tended to concentrate attention on individual organisms rather than on the biological continuum. The origin of life is thus looked for as a unique event in which an organism arises from the surrounding milieu. A more ecologically balanced point of view would examine the proto-ecological cycles and subsequent chemical systems that must have developed and flourished while objects resembling organisms appeared.[7]

No individual organism can exist in isolation. Animals depend on the photosynthesis of plants for their energy needs; plants depend on the carbon dioxide produced by animals, as well as on the nitrogen fixed by the bacteria at their roots; and together plants, animals and microorganisms regulate the entire biosphere and maintain the conditions conducive to life. According to the Gaia theory of James Lovelock and Lynn Margulis,[8] the evolution of the first living organisms went hand in hand with the transformation of the planetary surface from an inorganic environment to a self-regulating biosphere. "In that sense," writes Harold Morowitz, "life is a property of planets rather than of individual organisms."[9]

Life Defined in Terms of DNA

Let us now return to the question "What is life?" and ask: How does a bacterial cell work? What are its defining characteristics? When we look at a cell under an electron microscope, we notice that its metabolic processes involve special macromolecules—very large molecules consisting of long chains of hundreds of atoms. Two kinds of these macromolecules are found in all cells: proteins and nucleic acids (DNA and RNA).

In the bacterial cell, there are essentially two types of proteins—enzymes, which act as catalysts of various metabolic processes, and structural proteins, which are part of the cell structure. In higher organisms, there are also many other types of proteins with specialized functions, such as the antibodies of the immune system or the hormones.

Since most metabolic processes are catalyzed by enzymes and enzymes are specified by genes, the cellular processes are genetically controlled, which gives them great stability. The RNA molecules serve as messengers, delivering coded information for the synthesis of enzymes from the DNA, thus establishing the critical link between the cell's genetic and metabolic features.

DNA is also responsible for the cell's self-replication, which is a crucial characteristic of life. Without it, any accidentally formed structures would have decayed and disappeared, and life could never have evolved. This overriding importance of DNA might suggest that it should be identified as *the* single defining characteristic of life. We might simply say: "Living systems are chemical systems that contain DNA."

The problem with this definition is that dead cells also contain DNA. Indeed, DNA molecules may be preserved for hundreds, even thousands, of years after the organism dies. A spectacular example of such a case was reported a few years ago, when scientists in Germany succeeded in identifying the precise gene sequence in DNA from a Neanderthal skull—bones that had been dead for over 100,000 years![10] Thus, the presence of DNA alone is not sufficient to define life. At the very least, our definition would have to be modified to: "Living systems are chemical systems that contain DNA, and which are not dead." But then we would be saying, essentially, "a living system is a system that is alive"—a mere tautology.

This little exercise shows us that the molecular structures of the cell are not sufficient for the definition of life. We also need to describe the cell's metabolic processes—in other words, the patterns of relationships between the macromolecules. In this approach, we focus on the cell as a whole rather than on its parts. According to biochemist Pier Luigi Luisi, whose special field of research is molecular evolution and the origin of life, these two approaches—the "DNA-centered" view and the "cell-centered" view—represent two main philosophical and experimental streams in life sciences today.[11]

Membranes—The Foundation of Cellular Identity

Let us now look at the cell as a whole. A cell is characterized, first of all, by a boundary (the cell membrane) which discriminates between the system—the "self," as it were—and its environment. Within this boundary, there is a network of chemical reactions (the cell's metabolism) by which the system sustains itself.

Most cells have other boundaries besides membranes, such as rigid cell walls or capsules. These are common features in many kinds of cells, but only membranes are a universal feature of cellular life. Since its beginning, life on Earth has been associated with water. Bacteria move in water, and the metabolism inside their membranes takes place in a watery environment. In such fluid surroundings, a cell could never persist as a distinct entity without a physical barrier against free diffusion. The existence of membranes is therefore an essential condition for cellular life. Membranes are not only a universal characteristic of life, but also display the same type of structure throughout the living world. We shall see that the molecular details of this universal membrane structure hold important clues about the origin of life.[12]

A membrane is very different from a cell wall. Whereas cell walls are rigid structures, membranes are always active, opening and closing continually, keeping certain substances out and letting others in. The cell's metabolic reactions involve a variety of ions,* and the membrane, by being semipermeable, controls their proportions and keeps them in balance. Another critical activity of the membrane is to continually pump out excessive calcium waste, so that the calcium remaining within the cell is kept at the precise, very low level required for its metabolic functions. All these activities help to maintain the cell as a distinct entity and protect it from harmful environmental influences. Indeed, the first thing a bacterium does when it is attacked by another organism is to make membranes.[13]

All nucleated cells, and even most bacteria, also have internal membranes. In textbooks, a plant or animal cell is usually pictured as a large disk, surrounded by the cell membrane and containing a number of smaller disks (the organelles), each surrounded by its own membrane.[14] This picture is not really accurate. The cell does not contain several distinct membranes, but rather has one single, interconnected membrane system. This so-called "endomembrane system" is always in motion, wrapping itself around all the organelles and going out to the edge of the cell. It is a moving "conveyor belt" that is continually produced, broken down and produced again.[15]

*Ions are atoms that have net electric charge as a result of having lost or gained one or more electrons.

Through its various activities the cellular membrane regulates the cell's molecular composition and thus preserves its identity. There is an interesting parallel here to recent thinking in immunology. Some immunologists now believe that the central role of the immune system is to control and regulate the molecular repertoire throughout the organism, thus maintaining the organism's "molecular identity."[16] At the cellular level, the cell membrane plays a similar role. It regulates molecular compositions and, in doing so, maintains the cellular identity.

Self-generation

The cell membrane is the first defining characteristic of cellular life. The second characteristic is the nature of the metabolism that takes place within the cell boundary. In the words of microbiologist Lynn Margulis: "Metabolism, the incessant chemistry of self-maintenance, is an essential feature of life . . . Through ceaseless metabolism, through chemical and energy flow, life continuously produces, repairs, and perpetuates itself. Only cells, and organisms composed of cells, metabolize."[17]

When we take a closer look at the processes of metabolism, we notice that they form a chemical network. This is another fundamental feature of life. As ecosystems are understood in terms of food webs (networks of organisms), so organisms are viewed as networks of cells, organs and organ systems, and cells as networks of molecules. One of the key insights of the systems approach has been the realization that the network is a pattern that is common to all life. Wherever we see life, we see networks.

The metabolic network of a cell involves very special dynamics that differ strikingly from the cell's nonliving environment. Taking in nutrients from the outside world, the cell sustains itself by means of a network of chemical reactions that take place inside the boundary and produce all of the cell's components, including those of the boundary itself.[18]

The function of each component in this network is to transform or replace other components, so that the entire network continually generates itself. This is the key to the systemic definition of life: living networks continually create, or re-create, themselves by transforming or replacing their components. In this way they undergo continual structural changes while preserving their weblike patterns of organization.

The dynamic of self-generation was identified as a key characteristic of life by biologists Humberto Maturana and Francisco Varela, who gave it the name "autopoiesis" (literally, "self-making").[19] The concept of autopoiesis combines the two defining characteristics of cellular life mentioned above, the physical boundary and the metabolic network. Unlike the surfaces of crystals or large molecules, the boundary of an autopoietic system is chemically distinct from the rest of the system, and it participates in metabolic processes by assembling itself and by selectively filtering incoming and outgoing molecules.[20]

The definition of a living system as an autopoietic network means that the phenomenon of life has to be understood as a property of the system as a whole. In the words of Pier Luigi Luisi, "Life cannot be ascribed to any single molecular component (not even DNA or RNA!) but only to the entire bounded metabolic network."[21]

Autopoiesis provides a clear and powerful criterion for distinguishing between living and nonliving systems. For example, it tells us that viruses are not alive, because they lack their own metabolism. Outside living cells, viruses are inert molecular structures consisting of proteins and nucleic acids. A virus is essentially a chemical message that needs the metabolism of a living host cell to produce new virus particles, according to the instructions encoded in its DNA or RNA. The new particles are not built within the boundary of the virus itself, but outside in the host cell.[22]

Similarly, a robot that assembles other robots out of parts that are built by some other machines cannot be considered living. In recent years, it has often been suggested that computers and other automata may constitute future life-forms. However, unless they were able to synthesize their components from "food molecules" in their environment, they could not be considered to be alive according to our definition of life.[23]

The Cellular Network

As soon as we begin to describe the metabolic network of a cell in detail, we see that it is very complex indeed, even for the simplest bacteria. Most metabolic processes are facilitated (catalyzed) by enzymes and receive energy through special phosphate molecules known as ATP. The enzymes alone form an intricate network of catalytic reactions, and the ATP molecules form a corresponding energy network.[24] Through the messenger RNA, both of these networks are linked to the genome (the cell's DNA molecules), which is itself a complex interconnected web, rich in feedback loops, in which genes directly and indirectly regulate each other's activity.

Some biologists distinguish between two types of production processes and, accordingly, between two distinct cellular networks. The first is called, in a more technical sense of the term, the "metabolic" network, in which the "food" that enters through the cell membrane is turned into the so-called "metabolites"—the building blocks out of which the macromolecules—the enzymes, structural proteins, RNA, and DNA—are formed.

The second network involves the production of the macromolecules from the metabolites. This network includes the genetic level but extends to levels beyond the genes, and is therefore known as the "epigenetic"* network. Although these two networks have been given different names, they are closely interconnected and together form the autopoietic cellular network.

A key insight of the new understanding of life has been that biological forms and functions are not simply determined by a genetic blueprint but are emergent properties of the entire epigenetic network. To understand their emergence, we need to understand not only

*From the Greek *epi* ("above" or "beside").

the genetic structures and the cell's biochemistry, but also the complex dynamics that unfold when the epigenetic network encounters the physical and chemical constraints of its environment.

According to nonlinear dynamics, the new mathematics of complexity, this encounter will result in a limited number of possible functions and forms, described mathematically by attractors—complex geometric patterns that represent the system's dynamic properties.[25] Biologist Brian Goodwin and mathematician Ian Stewart have taken important first steps in using nonlinear dynamics to explain the emergence of biological form.[26] According to Stewart, this will be one of the most fruitful areas of science in the years to come:

> I predict—and I am by no means alone—that one of the most exciting growth areas of twenty-first-century science will be biomathematics. The next century will witness an explosion of new mathematical concepts, of new *kinds* of mathematics, brought into being by the need to understand the patterns of the living world.[27]

This view is quite different from the genetic determinism that is still very widespread among molecular biologists, biotechnology companies and in the popular scientific press.[28] Most people tend to believe that biological form is determined by a genetic blueprint, and that all the information about cellular processes is passed on to the next generation through the DNA when a cell divides and its DNA replicates. This is not at all what happens.

When a cell reproduces, it passes on not only its genes, but also its membranes, enzymes, organelles—in short, the whole cellular network. The new cell is not produced from naked DNA, but from an unbroken continuation of the entire autopoietic network. Naked DNA is never passed on, because genes can only function when they are embedded in the epigenetic network. Thus life has unfolded for over three billion years in an uninterrupted process, without ever breaking the basic pattern of its self-generating networks.

Emergence of New Order

The theory of autopoiesis identifies the pattern of self-generating networks as a defining characteristic of life, but it does not provide a detailed description of the physics and chemistry that are involved in these networks. As we have seen, such a description is crucial to understanding the emergence of biological forms and functions.

The starting point for this is the observation that all cellular structures exist far from thermodynamic equilibrium and would soon decay toward the equilibrium state—in other words, the cell would die—if the cellular metabolism did not use a continual flow of energy to restore structures as fast as they are decaying. This means that we need to describe the cell as an open system. Living systems are organizationally closed—they are autopoietic networks—but materially and energetically open. They need to feed on continual flows of matter and energy from their environment to stay alive. Conversely, cells, like all living organisms, continually produce waste, and this flow-through of matter—food and waste—establishes their place in the food web. In the words of Lynn Margulis, "The cell has an automatic relationship with somebody else. It leaks something, and somebody else will eat it."[29]

Detailed studies of the flow of matter and energy through complex systems have resulted in the theory of dissipative structures developed by Ilya Prigogine and his collaborators.[30] A dissipative structure, as described by Prigogine, is an open system that maintains itself in a state far from equilibrium, yet is nevertheless stable: the same overall structure is maintained in spite of an ongoing flow and change of components. Prigogine chose the term "dissipative structures" to emphasize this close interplay between structure on the one hand and flow and change (or dissipation) on the other.

The dynamics of these dissipative structures specifically include the spontaneous emergence of new forms of order. When the flow of energy increases, the system may encounter a point of instability, known as a "bifurcation point," at which it can branch off into an entirely new state where new structures and new forms of order may emerge.

This spontaneous emergence of order at critical points of instability is one of the most important concepts of the new understanding of life. It is technically known as self-organization and is often referred to simply as "emergence." It has been recognized as the dynamic origin of development, learning and evolution. In other words, creativity—the generation of new forms—is a key property of all living systems. And since emergence is an integral part of the dynamics of open systems, we reach the important conclusion that open systems develop and evolve. Life constantly reaches out into novelty.

The theory of dissipative structures, formulated in terms of nonlinear dynamics, explains not only the spontaneous emergence of order, but also helps us to define complexity.[31] Whereas traditionally the study of complexity has been a study of complex structures, the focus is now shifting from the structures to the processes of their emergence. For example, instead of defining the complexity of an organism in terms of the number of its different cell types, as biologists often do, we can define it as the number of bifurcations the embryo goes through in the organism's development. Accordingly, Brian Goodwin speaks of "morphological complexity."[32]

Prebiotic Evolution

Let us pause for a moment to review the defining characteristics of living systems that we have identified in our discussion of cellular life. We have learned that a cell is a membrane-bounded, self-generating, organizationally closed metabolic network; that it is materially and energetically open, using a constant flow of matter and energy to produce, repair and perpetuate itself; and that it operates far from equilibrium, where new structures and new forms of order may spontaneously emerge, thus leading to development and evolution. These characteristics are described by two different theories, representing two different perspectives on life—the theory of autopoiesis and the theory of dissipative structures.

When we try to integrate these two theories, we discover that there is a certain mismatch. While all autopoietic systems are dissipative structures, not all dissipative structures are autopoietic systems. Ilya Prigogine developed his theory from the study of complex thermal systems and chemical cycles that exist far from equilibrium, even though he was motivated to do so by a keen interest in the nature of life.[33]

Dissipative structures, then, are not necessarily living systems, but since emergence is an integral part of their dynamics, all dissipative structures have the potential to evolve. In other words, there is a "prebiotic" evolution—an evolution of inanimate matter that must have begun some time before the emergence of living cells. This view is widely accepted among scientists today.

The first comprehensive version of the idea that living matter originated from inanimate matter by a continuous evolutionary process was introduced into science by the Russian biochemist Alexander Oparin in his classic book *Origin of Life*, published in 1929.[34] Oparin called it "molecular evolution," and today it is commonly referred to as "prebiotic evolution." In the words of Pier Luigi Luisi, "Starting from small molecules, compounds with increasing molecular complexity and with emergent novel properties would have evolved, until the most extraordinary of emergent properties—life itself—originated."[35]

Although the idea of prebiotic evolution is now widely accepted, there is no consensus among scientists about the details of this process. Several scenarios have been proposed, but none have been demonstrated. One scenario begins with catalytic cycles and "hypercycles" (cycles of multiple feedback loops) formed by enzymes, which are capable of self-replication and evolution.[36] A different scenario is based on the recent discovery that certain kinds of RNA can also act as enzymes, i.e. as catalysts of metabolic processes. This catalytic ability of RNA, which is now well established, makes it possible to imagine an evolutionary stage in which two functions that are crucial to the living cell—information transfer and catalytic activities—were combined in a single type of molecule. Scientists have called this hypothetical stage the "RNA world."[37]

In the evolutionary scenario of the RNA world[38] the RNA molecules would first perform the catalytic activities necessary to assemble copies of themselves and would then begin to synthesize proteins, including enzymes. These newly built enzymes would be much more effective catalysts than their RNA counterparts and would eventually dominate. Finally, DNA would appear on the scene as the ultimate carrier of genetic information, with the added ability to correct transcription errors because of its double-stranded structure. At this stage, RNA would be relegated to the intermediary role it has today, displaced by DNA for more effective information storage and by protein enzymes for more effective catalysis.

Minimal Life

All these scenarios are still very speculative, whether they feature catalytic hypercycles of proteins (enzymes) surrounding themselves with membranes and then, somehow, creating a DNA structure, an RNA world evolving into today's DNA plus RNA plus proteins, or a synthesis of these two scenarios, which has recently been proposed.[39] No matter what the scenario of prebiotic evolution, the interesting question arises of whether we can talk about living systems at some stage before the appearance of cells. In other words, is there a way to define minimal features of living systems that may have existed in the past, irrespective of what has subsequently evolved? Here is the answer given by Luisi:

> It is clear that the process leading to life is a continuum process, and this makes an unequivocal definition of life very difficult. In fact, there are obviously many places in Oparin's pathway where the marker "minimal life" could arbitrarily be placed: at the level of self-replication; at the stage where self-replication was . . . accompanied by chemical evolution; at the point in time when proteins and nucleic

acids began to interact; when a genetic code was formed, or when the first cell was formed.[40]

Luisi comes to the conclusion that different definitions of minimal life, although equally justifiable, may be more or less meaningful depending on the purpose for which they are used.

If the basic idea of prebiotic evolution is correct, it should be possible, in principle, to demonstrate it in the laboratory. The challenge for scientists working in this field is to build life from molecules or, at least, to reconstruct different evolutionary steps in various prebiotic scenarios. Since there is no fossil record of evolving prebiotic systems from the time when the first rocks were formed on Earth to the emergence of the first cell, chemists have no helpful clues about possible intermediate structures, and their challenge might seem overwhelming.

Nevertheless, significant progress has been made recently, and we should also remember that this field is still very young. Systematic research into the origin of life has not been pursued for more than forty or fifty years, but even though our detailed ideas about prebiotic evolution are still very speculative, most biologists and biochemists do not doubt that life originated on Earth as the result of a sequence of chemical events, subject to the laws of physics and chemistry and to the nonlinear dynamics of complex systems.

This point is argued eloquently and in impressive detail by Harold Morowitz in a wonderful little book, *Beginnings of Cellular Life*,[41] which I shall follow closely for the remainder of this chapter. Morowitz approaches the question of prebiotic evolution and the origin of life from two sides. First, he identifies the basic principles of biochemistry and molecular biology that are common to all living cells. He traces these principles back through evolution to the origin of bacterial cells and argues that they must have played a major role in the formation of the "protocells" from which the first cells evolved: "Because of historical continuity, prebiotic processes should leave a signature in contemporary biochemistry."[42]

Having identified the basic principles of physics and chemistry that must have operated in the formation of protocells, Morowitz then asks: how could matter, subject to these principles and to the energy flows that were available on the surface of the Earth, have organized itself so as to bring forth various stages of protocells and then, eventually, the first living cell?

The Elements of Life

The basic elements of the chemistry of life are its atoms, molecules and chemical processes, or "metabolic pathways." In his detailed discussion of these elements, Morowitz shows beautifully that the roots of life reach deep into basic physics and chemistry.

We can start from the observation that multiple chemical bonds are essential to the formation of complex biochemical structures, and that carbon (C), nitrogen (N) and oxygen (O) are the only atoms that regularly form multiple bonds. We know that light elements make the strongest chemical bonds. It is therefore not surprising that these three elements, together with the lightest element, hydrogen (H), are the major atoms of biological structure.

We also know that life began in water and that cellular life still functions in a watery environment. Morowitz points out that water molecules (H_2O) are electrically highly polar, because their electrons stay closer to the oxygen atom than to the hydrogen atoms, so that they leave an effective positive charge on the H and a negative charge on the O. This polarity is a key feature in the molecular details of biochemistry and particularly in the formation of membranes, as we shall see below.

The last two major atoms of biological systems are phosphorus (P) and sulphur (S). These elements have unique chemical characteristics because of the great versatility of their compounds, and biochemists believe that they must have been major components of prebiotic chemistry. In particular, certain phosphates are instrumental in transforming and distributing chemical energy, which was as critical in prebiotic evolution as it is today in all cellular metabolism.

Moving on from atoms to molecules, there is a universal set of small organic molecules that is used by all cells as food for their metabolism. Although animals ingest many large and complex molecules, they are always broken down into small components before they enter into the metabolic processes of the cells. Moreover, the total number of different food molecules is not more than a few hundred, which is remarkable in view of the fact that an enormous number of small compounds can be made from the atoms of C, H, N, O, P and S.

The universality and small number of types of atoms and molecules in contemporary living cells is a strong indication of their common evolutionary origin in the first protocells, and this hypothesis is strengthened further when we turn to the metabolic pathways that constitute the basic chemistry of life. Once more, we encounter the same phenomenon. In the words of Morowitz: "Amid the enormous diversity of biological types, including millions of recognizable species, the variety of biochemical pathways is small, restricted, and universally distributed."[43] It is very likely that the core of this metabolic network, or "metabolic chart," represents a primordial biochemistry that holds important clues about the origin of life.

Bubbles of Minimal Life

As we have seen, the careful observation and analysis of the basic elements of life strongly suggests that cellular life is rooted in a universal physics and biochemistry, which existed long before the evolution of living cells. Let us now turn to the second line of investigation presented by Harold Morowitz. How could matter have organized itself within the constraints of that primordial physics and biochemistry, without any extra ingredients, so as to evolve into the complex molecules from which life emerged?

The idea that small molecules in a primordial "chemical soup" should assemble spontaneously into structures of ever-increasing complexity runs counter to all conventional experience with simple chemical systems. Many scientists have therefore argued that the odds of such a prebiotic evolution are vanishingly small; or, alternatively, that there must have been an extraordinary triggering event, such as a seeding of the Earth with macromolecules by meteorites.

Today, our starting position for resolving this puzzle is radically different. Scientists working in this field have come to recognize that the flaw of the conventional argument lies in the idea that life must have emerged out of a primordial chemical soup through a progressive increase in molecular complexity. The new thinking, as Morowitz emphasizes repeatedly, begins from the hypothesis that very early on, before the increase of molecular complexity, certain molecules assembled into primitive membranes that spontaneously formed closed bubbles, and that the evolution of molecular complexity took place inside these bubbles, rather than in a structureless chemical soup.

Before going into the details of how primitive membrane-bounded bubbles, known to chemists as "vesicles," could have formed spontaneously, I want to discuss the dramatic consequences of such a process. With the formation of vesicles two different environments—an outside and an inside—were established, in which compositional differences could develop.

As Morowitz shows, the internal volume of a vesicle provides a closed microenvironment in which directed chemical reactions can occur, which means that molecules that are normally rare may be formed in great quantities. These molecules include in particular the building blocks of the membrane itself, which become incorporated into the existing membrane, so that the whole membrane area increases. At some point in this growth process the stabilizing forces are no longer able to maintain the membrane's integrity, and the vesicle breaks up into two or more smaller bubbles.[44]

These processes of growth and replication will occur only if there is a flow of energy and matter through the membrane. Morowitz describes plausibly how this might have happened.[45] The vesicle membranes are semipermeable, and thus various small molecules can enter the bubbles or be incorporated into the membrane. Among those will be chromophores, molecules that absorb sunlight. Their presence creates electric potentials across the membrane, and thus the vesicle becomes a device that converts light energy into electric potential energy.

Once this system of energy conversion is in place, it becomes possible for a continuous flow of energy to drive the chemical processes inside the vesicle. Eventually, a further refinement of this energy scenario takes place when the chemical reactions in the bubbles produce phosphates, which are very effective in the transformation and distribution of chemical energy.

Morowitz also points out that the flow of energy and matter is necessary not only for the growth and replication of vesicles, but also for the mere persistence of stable structures. Since all such structures arise from chance events in the chemical domain and are subject to thermal decay, they are by their very nature not in equilibrium and can only be preserved through continual processing of matter and energy.[46] At this point it becomes apparent that two defining characteristics of cellular life are manifest in rudimentary form in these primitive membrane-bounded bubbles. The vesicles are open systems, subject to continual flows of energy and matter, while their interiors are relatively closed spaces in which networks of chemical reactions are likely to develop. We can recognize these two properties as the roots of living networks and their dissipative structures.

Now the stage is set for prebiotic evolution. In a large population of vesicles there will be many differences in their chemical properties and structural components. If these differences persist when the bubbles divide, we can speak of a pregenetic memory and of species of vesicles, and since these species will compete for energy and various molecules from their environment, a kind of Darwinian dynamic of competition and natural selection will take place, in which molecular accidents may be amplified and selected for their "evolutionary" advantages. In addition, different types of vesicles will occasionally fuse, which may result in synergies of advantageous chemical properties, foreshadowing the phenomenon of symbiogenesis (the creation of new forms of life through the symbiosis of the organisms) in biological evolution.[47]

Thus we see that a variety of purely physical and chemical mechanisms provides the membrane-bounded vesicles with the potential to evolve through natural selection into complex, self-producing structures without enzymes or genes in these early stages.[48]

Membranes

Let us now return to the formation of membranes and membrane-bounded bubbles. According to Morowitz, the formation of these bubbles constitutes the most crucial step in prebiotic evolution: "It is the closure of [a primitive] membrane into a 'vesicle' that represents a discrete transition from nonlife to life."[49]

The chemistry of this crucial process is surprisingly simple and common. It is based on the electric polarity of water mentioned above. Because of this polarity, certain molecules are hydrophilic (attracted by water), while others are hydrophobic (repelled by water). A third kind of molecules are those of fatty and oily substances, known as lipids. They are elongated structures with one hydrophilic and one hydrophobic end, as pictured below.

hydrophobic end ▭—○ hydrophilic end

Lipid molecule, adapted from Morowitz (1992).

When these lipids come in contact with water, they spontaneously form a variety of structures. For example, they may form a monomolecular film spreading over the water surface (see Figure A), or they may coat oil droplets and keep them suspended in water (see Figure B). Such coating of oil occurs in mayonnaise and also accounts for the action of soaps in removing oil stains. Alternatively, the lipids may coat water droplets for suspension in oil (see Figure C).

The lipids may form an even more complex structure, consisting of a double layer of molecules with water on both sides, as shown in Figure D. This is the basic membrane structure, and just like the single layer of molecules, it too may form droplets, which are the membrane-bounded vesicles under discussion (see Figure E). These double-layered greasy membranes show a surprising number of properties that are quite similar to contemporary cellular membranes. They restrict the number of molecules that can enter the vesicle, transform solar energy into electrical energy and even collect phosphate compounds in-

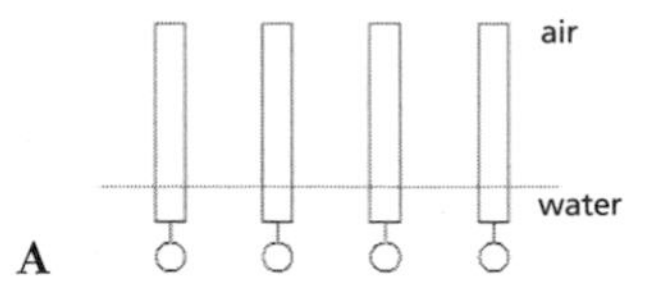

monomolecular film on water surface

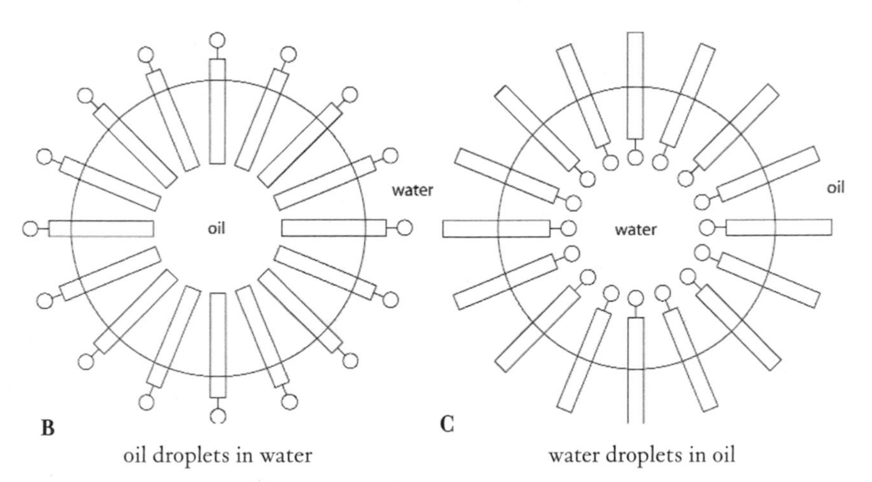

oil droplets in water

water droplets in oil

Simple structures formed by lipid molecules, adapted from Morowitz (1992).

side their structure. Indeed, today's cellular membranes seem to be a refinement of the primordial membranes. They too consist mainly of lipids with proteins attached or inserted into the membrane.

Lipid vesicles, then, are the ideal candidates for the protocells out of which the first living cells evolved. As Morowitz reminds us, their properties are so astonishing that it is important not to forget that they are structures that form spontaneously according to the basic laws of physics and chemistry.[50] They form as naturally as bubbles when you put oil and water together and shake the mixture.

In the scenario outlined by Morowitz, the first protocells formed around 3.9 billion years ago when the planet had cooled down, shallow oceans and the first rocks had been formed, and carbon had combined with the other fundamental elements of life to form a great variety of chemical compounds.

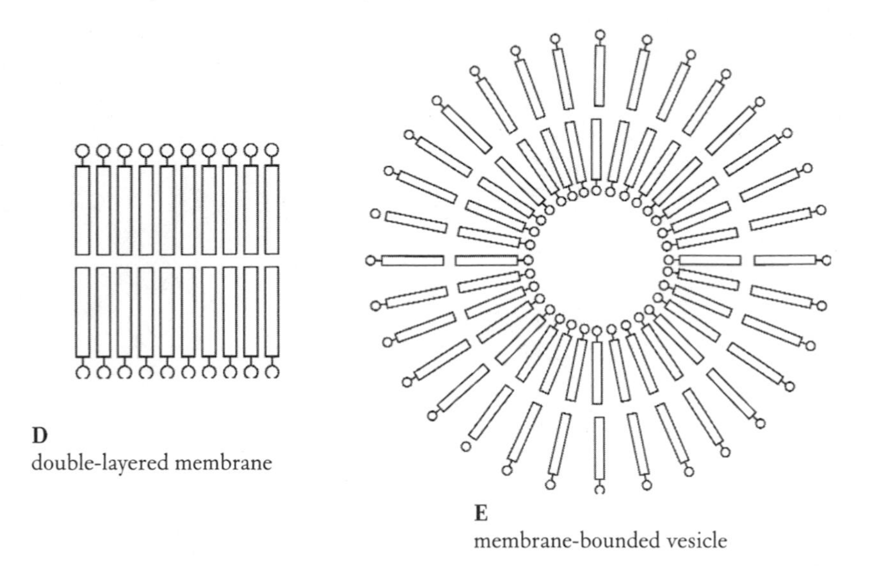

Membrane and vesicle formed by lipid molecules, adapted from Morowitz (1992).

Among these compounds were oily substances called paraffins, which are long hydrocarbon chains. The interactions of these paraffins with water and various dissolved minerals led to the lipids; these in turn condensed to a variety of droplets and also formed thin, single-layered and double-layered sheets. Under the influence of wave action, the sheets spontaneously formed closed vesicles, and thus began the transition of life.

Re-creating Protocells in the Laboratory

This scenario is still highly speculative, because so far chemists have not been able to produce lipids from small molecules. All the lipids in our environment are derived from petroleum and other organic substances. However, focusing on membranes and vesicles rather than on DNA and RNA has given rise to an exciting new direction of research that has already produced many encouraging results.

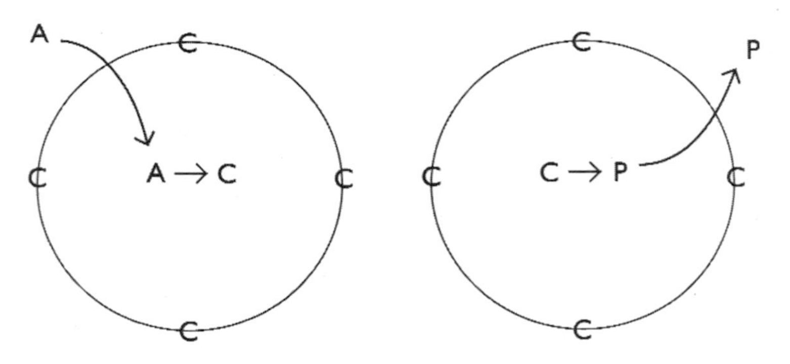

The two basic reactions in a minimal autopoietic system, from Luisi (1993).

One of the pioneering research teams in this field is led by Pier Luigi Luisi at the Swiss Federal Institute of Technology (ETH) in Zurich. Luisi and his colleagues succeeded in preparing simple "soap and water" environments in which vesicles of the type described above form spontaneously and, depending on the chemical reactions involved, perpetuate themselves, grow and self-replicate or collapse again.[51]

Luisi has emphasized that the self-replicating vesicles produced in his laboratory are minimal autopoietic systems in which chemical reactions are enclosed by a boundary assembled from the very products of the reactions. In the simplest case, illustrated above, the boundary is composed of only one component, C. There is only one type of molecule, A, that can enter through the membrane and generate C in the reaction A → C inside the bubble. In addition, there is a decomposition reaction, C → P, and the product P leaves the vesicle. Depending on the relative rates of these two basic reactions, the vesicle will either grow and self-replicate, remain stable or collapse.

Luisi and his colleagues have carried out experiments with vesicles of many types and have tested a variety of chemical reactions taking place inside these bubbles.[52] By producing spontaneously formed autopoietic protocells, these biochemists have re-created what was perhaps the most critical step in prebiotic evolution.

Catalysts and Complexity

Once the protocells were formed and the molecules for absorption and transformation of solar energy were in place, the evolution toward greater complexity could begin. At this stage, the elements of the chemical compounds were C, H, O, P, and possibly S. With the entry of nitrogen into the system, probably in the form of ammonia (NH_3), a dramatic increase in molecular complexity became possible, because nitrogen is essential for two characteristic features of cellular life—catalysis and information storage.[53]

Catalysts increase the rates of chemical reactions without being changed themselves in the process, and they make possible reactions that could not occur without them. Catalytic reactions are crucial processes in the chemistry of life. In contemporary cells they are mediated by enzymes, but in the early stages of protocells these elaborate macromolecules did not exist.

However, chemists have discovered that certain small molecules that bond to membranes may also have catalytic properties. Morowitz assumes that the entry of nitrogen into the chemistry of the protocells led to the formation of such primitive catalysts. In the meantime, the biochemists at ETH have succeeded in re-creating this evolutionary step by attaching molecules with weak catalytic properties to the membranes of the vesicles formed in their laboratory.[54]

With the appearance of catalysts molecular complexity increased rapidly, because catalysts create chemical networks by interlinking different reactions. Once this happens, the entire nonlinear dynamics of networks come into play. This includes in particular the spontaneous emergence of new forms of order, as demonstrated by Ilya Prigogine and Manfred Eigen, two Nobel laureates in chemistry who pioneered the study of self-organizing chemical systems.[55]

With the help of catalytic reactions, beneficial chance events would have been enhanced considerably, and thus a fully Darwinian mode of competition would have developed, constantly pushing the protocells toward increasing complexity, further from equilibrium and closer to life.

The final step in the emergence of life from protocells was the evolution of proteins, nucleic acids and the genetic code. At present, the details of this stage are still quite mysterious, but we need to remember that the evolution of catalytic networks within the closed spaces of the protocells created a new type of network chemistry that is still very poorly understood. We can expect that the application of nonlinear dynamics to these complex chemical networks, as well as the "explosion of new mathematical concepts" predicted by Ian Stewart, will shed considerable light on the last phase of prebiotic evolution. Harold Morowitz points out that the analysis of the chemical pathways from small molecules to amino acids reveals an extraordinary set of correlations that seem to suggest a "deep network logic" in the development of the genetic code.[56]

Another interesting discovery is that chemical networks in closed spaces that are subject to continual flows of energy develop processes surprisingly like those of ecosystems. For example, significant features of biological photosynthesis and the ecological carbon cycle have been shown to emerge in laboratory systems. The cycling of matter seems to be a general feature of chemical networks that are kept far from equilibrium by a constant flux of energy.[57]

"An abiding message," Morowitz concludes, "is the necessity of understanding the complex network of organic reactions containing intermediates that are catalytic for other reactions . . . If we better understood how to deal with chemical networks, many other problems in prebiotic chemistry would become appreciably simpler."[58] When more biochemists become interested in nonlinear dynamics, it is likely that the new "biomathematics" envisaged by Stewart will include a proper theory of chemical networks, and that this new theory will finally reveal the secrets of the last stage in the emergence of life.

The Unfolding of Life

Once memory became encoded in macromolecules, the membrane-bounded chemical networks acquired all the essential characteristics of today's bacterial cells. This major signpost in the evolution of life established itself perhaps 3.8 billion years ago, about 100 million years after the formation of the first protocells. This marked the emergence of a universal ancestor—either a single cell or a population of cells—from which all subsequent life on Earth descended. As Morowitz explains: "Although we do not know how many independent origins of cellular life may have occurred, all present life is descended from a single clone. This follows from the universality of the basic biochemical networks and programmes of macromolecular synthesis."[59] This universal ancestor must have outperformed all the protocells. Thus its descendants took over the Earth, weaving a planetary bacterial web and occupying all the ecological niches, so that the emergence of other forms of life became impossible.

The global unfolding of life proceeded through three major avenues of evolution.[60] The first, but perhaps least important, is the random mutation of genes, the centerpiece of neo-Darwinian theory. Gene mutation is caused by a chance error in the self-replication of DNA, when the two chains of the DNA's double helix separate and each of them serves as a template for the construction of a new complementary chain. Those chance errors do not seem to occur frequently enough to explain the evolution of the great diversity of life-forms, given the well-known fact that most mutations are harmful and only very few result in useful variations.[61]

In the case of bacteria the situation is different, because bacteria divide so rapidly that billions of them can be generated from a single cell within days. Because of this enormous rate of reproduction, a single successful bacterial mutation can spread rapidly through its environment, and thus mutation is an important evolutionary avenue for bacteria.

Bacteria have also developed a second avenue of evolutionary creativity that is vastly more effective than random mutation. They freely pass hereditary traits from one to another in a global exchange network of incredible power and efficiency. The discovery of this global trading of genes, technically known as DNA recombination, must rank as one of the most astonishing discoveries of modern biology. Lynn Margulis describes it vividly: "Horizontal genetic transfer among bacteria is as if you jumped into a pool with brown eyes and came out with blue eyes."[62]

This gene transfer takes place continually, with many bacteria changing up to 15 percent of their genetic material on a daily basis. As Margulis explains, "When you threaten a bacterium, it will spill its DNA into the environment, and everyone around picks it up; and in a few months it will go all the way around the world."[63] Since all bacterial strains can potentially share hereditary traits in this way, some microbiologists argue that bacteria, strictly speaking, should not be classified into species.[64] In other words, all bacteria are part of a single microscopic web of life.

In evolution, then, bacteria are able rapidly to accumulate random mutations, as well as big chunks of DNA, through gene trading. Consequently, they have an astonishing ability to adapt to environmental changes. The speed with which drug resistance spreads among bacterial communities is dramatic proof of the efficiency of their communication networks. Microbiology teaches us the sobering lesson that technologies like genetic engineering and a global communications network, which are often considered to be advanced achievements of our modern civilization, have been used by the planetary web of bacteria for billions of years.

During the first two billion years of biological evolution, bacteria and other microorganisms were the only life forms on the planet. During those two billion years, bacteria continually transformed the Earth's surface and atmosphere, and established the global feedback loops for the self-regulation of the Gaia system. In so doing, they invented all of life's essential biotechnologies, including fermentation, photosynthesis, nitrogen fixation, respiration and various devices for rapid motion. Recent research in microbiology has made it evident that, as far as the processes of life are concerned, the planetary network of bacteria has been the main source of evolutionary creativity.

But what about the evolution of biological form, of the enormous variety of living beings in the visible world? If random mutations are not an effective evolutionary mechanism for them, and if they do not trade genes like bacteria, how have the higher forms of life evolved? This question was answered by Lynn Margulis with the discovery of a third avenue of evolution—evolution through symbiosis—that has profound implications for all branches of biology.

Symbiosis, the tendency of different organisms to live in close association with one another and often inside one another (like the bacteria in our intestines), is a widespread and well-known phenomenon. But Margulis went a step further and proposed the hypothesis that long-term symbioses involving bacteria and other microorganisms living inside larger cells have led and continue to lead to new forms of life. Margulis published her revolutionary hypothesis first in the mid-sixties, and over the years developed it into a full-fledged theory, now known as "symbiogenesis," which sees the creation of new forms of life through permanent symbiotic arrangements as the principal avenue of evolution for all higher organisms.[65]

Bacteria, again, have played a major role in this evolution through symbiosis. When certain small bacteria merged symbiotically with larger cells and continued to live inside them as organelles, the result was a giant step in evolution—the creation of plant and animal cells that reproduced sexually and eventually evolved into the living organisms we see in our environment. In their evolution, these organisms continued to absorb bacteria, incorporating parts of their genomes to synthesize proteins for new structures and new biological functions, not unlike the corporate mergers and acquisitions in today's business world. For example, evidence has been accumulating that the microtubules, which are essential to the architecture of the brain, were originally contributed by the "corkscrew" bacteria known as spirochetes.[66]

The evolutionary unfolding of life over billions of years is a breathtaking story, told beautifully by Lynn Margulis and Dorion Sagan in their book *Microcosmos*.[67] Driven by the creativity inherent in all living systems, expressed through the avenues of mutation, gene trading and symbiosis, and honed by natural selection, the planetary web of life expanded and complexified into forms of ever-increasing diversity.

This majestic unfolding did not proceed through continuous gradual changes over time. The fossil record shows clearly that throughout evolutionary history there have been long periods of stability, or stasis, without much genetic variation, punctuated by sudden and dramatic transitions.[68] This picture of "punctuated equilibria" indicates that the sudden transitions were caused by mechanisms quite different from the random mutations of neo-Darwinist theory, and the creation of new species through symbiosis seems to have played a critical role. As Margulis puts it, "From the long view of geological time, symbioses are like flashes of evolutionary lightning."[69]

Another striking pattern is the repeated occurrence of catastrophes followed by intense periods of growth and innovation. Thus, 245 million years ago, the most devastating mass extinctions the world has ever seen were rapidly followed by the evolution of mammals; and 66 million years ago the catastrophe that eliminated the dinosaurs from the face of the Earth cleared the way for the evolution of the first primates and, eventually, of the human species.

What Is Life?

Now, let us return to the question posed at the beginning of this chapter—What are the defining characteristics of living systems?—and summarize what we have learned. Focusing on bacteria as the simplest living systems, we characterized a living cell as a membrane-bounded, self-generating, organizationally closed metabolic network. This network involves several types of highly complex macromolecules: structural proteins; enzymes, which act as catalysts of metabolic processes; RNA, the messengers carrying genetic information; and DNA, which stores the genetic information and is responsible for the cell's self-replication.

We also learned that the cellular network is materially and energetically open, using a constant flow of matter and energy to produce, repair and perpetuate itself; and that it operates far from equilibrium, where new structures and new forms of order may spontaneously emerge, thus leading to development and evolution.

Finally, we have seen that a prebiotic form of evolution, involving membrane-enclosed bubbles of "minimal life," began long before the emergence of the first living cell; and that the roots of life reach deep into the basic physics and chemistry of these protocells.

We also identified three major avenues of evolutionary creativity—mutation, gene trading and symbiosis—through which life unfolded for over three billion years, from the universal bacterial ancestors to the emergence of human beings, without ever breaking the basic pattern of its self-generating networks.

To extend this understanding of the nature of life to the human social dimension, which is the central task of this book, we need to deal with conceptual thought, values, meaning and purpose—phenomena that belong to the realm of human consciousness and culture. This means that we need to include an understanding of mind and consciousness in our understanding of living systems.

As we shift our focus to the cognitive dimension of life, we shall see that a unified view of life, mind and consciousness is now emerging in which human consciousness is inextricably linked to the social world of interpersonal relationships and culture. Moreover, we shall discover that this unified view allows us to understand the spiritual dimension of life in a way that is fully consistent with traditional conceptions of spirituality.

| two |

MIND AND CONSCIOUSNESS

One of the most important philosophical implications of the new understanding of life is a novel conception of the nature of mind and consciousness, which finally overcomes the Cartesian division between mind and matter. In the seventeenth century, René Descartes based his view of nature on the fundamental division between two independent and separate realms—that of mind, the "thinking thing" (*res cogitans*), and that of matter, the "extended thing" (*res extensa*). This conceptual split between mind and matter has haunted Western science and philosophy for more than 300 years.

Following Descartes, scientists and philosophers continued to think of the mind as an intangible entity and were unable to imagine how this "thinking thing" is related to the body. Although neuroscientists have known since the nineteenth century that brain structures and mental functions are intimately connected, the exact relationship between mind and brain remained a mystery. As recently as 1994, the editors of an anthology titled *Consciousness in Philosophy and Cognitive Neuroscience* stated frankly in their introduction: "Even though everybody agrees that mind has something to do with the brain, there is still no general agreement on the exact nature of this relationship."[1]

The decisive advance of the systems view of life has been to abandon the Cartesian view of mind as a thing, and to realize that mind and consciousness are not things but processes. In biology, this novel concept of the mind was developed during the 1960s by Gregory Bateson, who used the term "mental process," and independently by Humberto Maturana, who focused on cognition, the process of knowing.[2] In the 1970s, Maturana and Francisco Varela expanded Maturana's initial work into a full theory, which has become known as the Santiago Theory of Cognition.[3] During the past twenty-five years, the study of the mind from this systemic perspective has blossomed into a rich interdisciplinary field, known as cognitive science, which transcends the traditional frameworks of biology, psychology and epistemology.

The Santiago Theory of Cognition

The central insight of the Santiago Theory is the identification of cognition, the process of knowing, with the process of life. Cognition, according to Maturana and Varela, is the activity involved in the self-generation and self-perpetuation of living networks. In other words, cognition is the very process of life. The organizing activity of living systems, at all levels of life, is mental activity. The interactions of a living organism—plant, animal or human—with its environment are cognitive interactions. Thus life and cognition are inseparably connected. Mind—or, more accurately, mental activity—is immanent in matter at all levels of life.

This is a radical expansion of the concept of cognition and, implicitly, the concept of mind. In this new view, cognition involves the entire process of life—including perception, emotion, and behavior—and does not even necessarily require a brain and a nervous system.

In the Santiago theory, cognition is closely linked to autopoiesis, the self-generation of living networks. The defining characteristic of an autopoietic system is that it undergoes continual structural changes while preserving its weblike pattern of organization. The components of the network continually produce and transform one another, and they do so in two distinct ways. One type of structural change is that of self-renewal. Every living organism continually renews itself, as its cells break down and build structures, and tissues and organs replace their cells in continual cycles. In spite of this ongoing change, the organism maintains its overall identity, or pattern of organization.

The second type of structural changes in a living system are those which create new structures—new connections in the autopoietic network. These changes, developmental rather than cyclical, also take place continually, either as a consequence of environmental influences or as a result of the system's internal dynamics.

According to the theory of autopoiesis, a living system couples to its environment structurally, i.e. through recurrent interactions, each of which triggers structural changes in the system. For example, a cell membrane continually incorporates substances from its environment into the cell's metabolic processes. An organism's nervous system changes its connectivity with every sense perception. These living systems are autonomous, however. The environment only triggers the structural changes; it does not specify or direct them.

Structural coupling, as defined by Maturana and Varela, establishes a clear difference between the ways living and nonliving systems interact with their environments. For example, when you kick a stone, it will *react* to the kick according to a linear chain of cause and effect. Its behavior can be calculated by applying the basic laws of Newtonian mechanics. When you kick a dog, the situation is quite different. The dog will respond with structural changes according to its own nature and (nonlinear) pattern of organization. The resulting behavior is generally unpredictable.

As a living organism responds to environmental influences with structural changes, these changes will in turn alter its future behavior. In other words, a structurally coupled system is a learning system. Continual structural changes in response to the environment—and consequently continuing adaptation, learning and development—are key characteristics of the behavior of all living beings. Because of its structural coupling, we can call the behavior of an animal intelligent but would not apply that term to the behavior of a rock.

As it keeps interacting with its environment, a living organism will undergo a sequence of structural changes, and over time it will form its own individual pathway of structural coupling. At any point on this pathway, the structure of the organism is a record of previous structural changes and thus of previous interactions. In other words, all living beings have a history. Living structure is always a record of prior development.

Now, since an organism records previous structural changes, and since each structural change influences the organism's future behavior, this implies that the behavior of the living organism is dictated by its structure. In Maturana's terminology, the behavior of living systems is "structure-determined."

This notion sheds new light on the age-old philosophical debate about freedom and determinism. According to Maturana, the behavior of a living organism is determined, but rather than being determined by outside forces, it is determined by the organism's own structure—a structure formed by a succession of autonomous structural changes. Hence the behavior of the living organism is both determined and free.

Living systems, then, respond autonomously to disturbances from the environment with structural changes, i.e. by rearranging their pattern of connectivity. According to Maturana and Varela, you can never direct a living system; you can only disturb it. More than that, the living system not only specifies its structural changes; it also specifies *which disturbances from the environment trigger them.* In other words, a living system maintains the freedom to decide what to notice and what will disturb it. This is the key to the Santiago Theory of Cognition. The structural changes in the system constitute acts of cognition. By specifying which perturbations from the environment trigger changes, the system specifies the extent of its cognitive domain; it "brings forth a world," as Maturana and Varela put it.

Cognition, then, is not a representation of an independently existing world, but rather a continual bringing forth of a world through the process of living. The interactions of a living system with its environment are cognitive interactions, and the process of living itself is a process of cognition. In the words of Maturana and Varela, "to live is to know." As a living organism goes through its individual pathway of structural changes, each of these changes corresponds to a cognitive act, which means that learning and development are merely two sides of the same coin.

The identification of mind, or cognition, with the process of life is a novel idea in science, but it is one of the deepest and most archaic intuitions of humanity. In ancient times, the rational human mind was seen as merely one aspect of the immaterial soul, or spirit. The basic distinction was not between body and mind, but between body and soul, or body and spirit.

In the languages of ancient times, both soul and spirit are described with the metaphor of the breath of life. The words for "soul" in Sanskrit (*atman*), Greek (*psyche*), and Latin (*anima*) all mean "breath." The same is true of the words for "spirit" in Latin (*spiritus*), Greek (*pneuma*), and Hebrew (*ruah*). These, too, mean "breath."

The common ancient idea behind all these words is that of soul or spirit as the breath of life. Similarly, the concept of cognition in the Santiago Theory goes far beyond the rational mind, as it includes the entire process of life. Describing cognition as the breath of life seems to be a perfect metaphor.

The conceptual advance of the Santiago Theory is best appreciated by revisiting the thorny question of the relationship between mind and brain. In the Santiago Theory, this relationship is simple and clear. The Cartesian characterization of mind as the "thinking thing" is abandoned. Mind is not a thing but a process—the process of cognition, which is identified with the process of life. The brain is a specific structure through which this process operates. The relationship between mind and brain, therefore, is one between process and structure. Moreover, the brain is not the only structure through which the process of cognition operates. The entire structure of the organism participates in the process of cognition, whether or not the organism has a brain and a higher nervous system.

In my view, the Santiago Theory of Cognition is the first scientific theory that overcomes the Cartesian division of mind and matter, and will thus have far-reaching implications. Mind and matter no longer appear to belong to two separate categories, but can be seen as representing two complementary aspects of the phenomenon of life—process and structure. At all levels of life, beginning with the simplest cell, mind and matter, process and structure, are inseparably connected.

Cognition and Consciousness

Cognition, as understood in the Santiago Theory, is associated with all levels of life and is thus a much broader phenomenon than consciousness. Consciousness—that is, conscious, lived experience—unfolds at certain levels of cognitive complexity that require a brain and a higher nervous system. In other words, consciousness is a special kind of cognitive process that emerges when cognition reaches a certain level of complexity.

It is interesting that the notion of consciousness as a process appeared in science as early as the late nineteenth century in the writings of William James, whom many consider the greatest American psychologist. James was a fervent critic of the reductionist and materialist theories that dominated psychology in his time, and an enthusiastic advocate of the interdependence of mind and body. He pointed out that consciousness is not a thing, but an ever-changing stream, and he emphasized the personal, continuous and highly integrated nature of this stream of consciousness.[4]

In subsequent years, however, the exceptional views of William James were not able to break the Cartesian spell on psychologists and natural scientists, and his influence did not reemerge until the last few decades of the twentieth century. Even during the 1970s and 1980s, when new humanistic and transpersonal approaches were formulated by American psychologists, the study of consciousness as lived experience was still taboo in cognitive science.

During the 1990s, the situation changed dramatically. While cognitive science established itself as a broad interdisciplinary field of study, new noninvasive techniques for analyzing brain functions were developed, which made it possible to observe complex neural processes associated with mental imagery and other human experiences.[5] And suddenly, the scientific study of consciousness became a respectable and lively field of research. Within a few years, several books about the nature of consciousness, authored by Nobel laureates and other eminent scientists, were published; dozens of articles by the leading cognitive scientists and philosophers appeared in the newly created *Journal of Consciousness Studies;* and "Toward a Science of Consciousness" became a popular theme for large scientific conferences.[6]

Although cognitive scientists and philosophers have proposed many different approaches to the study of consciousness, and have sometimes engaged in heated debates, it seems that there is a growing consensus on two important points. The first, as mentioned above, is the recognition that consciousness is a cognitive process, emerging from complex neural activity. The second point is the distinction between two types of consciousness—in other words, two types of cognitive experiences—which emerge at different levels of neural complexity.

The first type, known as "primary consciousness," arises when cognitive processes are accompanied by basic perceptual, sensory and emotional experience. Primary consciousness is probably experienced by most mammals and perhaps by some birds and other vertebrates.[7] The second type of consciousness, sometimes called "higher-order consciousness,"[8] involves self-awareness—a concept of self, held by a thinking and reflecting subject. This experience of self-awareness emerged during the evolution of the great apes, or "hominids," together with language, conceptual thought and all the other characteristics that fully unfolded in human consciousness. Because of the critical role of reflection in this higher-order conscious experience, I shall call it "reflective consciousness."

Reflective consciousness involves a level of cognitive abstraction that includes the ability to hold mental images, which allows us to formulate values, beliefs, goals and strategies. This evolutionary stage is of central relevance to the main theme of this book—the extension of the new understanding of life to the social domain—because with the evolution of language arose not only the inner world of concepts and ideas, but also the social world of organized relationships and culture.

The Nature of Conscious Experience

The central challenge of a science of consciousness is to explain the experience associated with cognitive events. Different states of conscious experience are sometimes called *qualia* by cognitive scientists, because each state is characterized by a special "qualitative feel."[9] The challenge of explaining these *qualia* has been called "the hard problem of consciousness" in an oft-cited article by the philosopher David Chalmers.[10] After reviewing conventional cognitive science, Chalmers asserts that it cannot explain why certain neural processes give rise to experience. "To account for conscious experience," he concludes, "we need an *extra ingredient* in the explanation."

This statement is reminiscent of the debate between mechanists and vitalists about the nature of biological phenomena during the early decades of the twentieth century.[11] Whereas the mechanists asserted that all biological phenomena can be explained in terms of the laws of physics and chemistry, the vitalists maintained that a "vital force" must be added to those laws as an additional, nonphysical "ingredient" to explain biological phenomena.

The insight that emerged from this debate, though not formulated until many decades later, is that in order to explain biological phenomena, we also need to take into account the complex nonlinear dynamics of living networks.

A full understanding of biological phenomena will be reached only when we approach it through the interplay of three different levels of description—the biology of the observed phenomena, the laws of physics and biochemistry, and the nonlinear dynamics of complex systems.

It seems to me that cognitive scientists find themselves in a very similar situation, albeit at a different level of complexity, when they approach the study of consciousness. Conscious experience is an emergent phenomenon, which means that it cannot be explained in terms of neural mechanisms alone. Experience emerges from the complex nonlinear dynamics of neural networks and can be explained only if our understanding of neurobiology is combined with an understanding of those dynamics.

To reach a full understanding of consciousness, we must approach it through the careful analysis of conscious experience; of the physics, biochemistry, and biology of the nervous system; and of the nonlinear dynamics of neural networks. A true science of consciousness will be formulated only when we understand how these three levels of description can be woven together into what Francisco Varela has called the "triple braid" of consciousness research.[12]

When the study of consciousness is approached by braiding together experience, neurobiology and nonlinear dynamics, the "hard problem" turns into the challenge of understanding and accepting two new scientific paradigms. The first is the paradigm of complexity theory. Since most scientists are used to working with linear models, they are often reluctant to adopt the nonlinear framework of complexity theory and find it difficult to appreciate fully the implications of nonlinear dynamics. This applies in particular to the phenomenon of emergence.

It seems quite mysterious that experience should emerge from neurophysiological processes. However, this is typical of emergent phenomena. Emergence results in the creation of novelty, and this novelty is often qualitatively different from the phenomena out of which it emerged. This can readily be illustrated with a well-known example from chemistry: the structure and properties of sugar.

When carbon, oxygen, and hydrogen atoms bond in a certain way to form sugar, the resulting compound has a sweet taste. The sweetness resides neither in the C, nor in the O, nor in the H; it resides in the pattern that emerges from their interaction. It is an emergent property. Moreover, strictly speaking, the sweetness is not a property of the chemical bonds. It is a sensory experience that arises when the sugar molecules interact with the chemistry of our taste buds, which in turn causes a set of neurons to fire in a certain way. The experience of sweetness emerges from that neural activity.

Thus, the simple statement that the characteristic property of sugar is its sweetness really refers to a series of emergent phenomena at different levels of complexity. Chemists have no conceptual problem with these emergent phenomena when they identify a certain class of compounds as sugars because of their sweet taste. Nor will future cognitive scientists have conceptual problems with other kinds of emergent phenomena when they analyze them in terms of the resulting conscious experience, as well as in terms of the relevant biochemistry and neurobiology.

To do so, however, scientists will need to accept another new paradigm—the recognition that the analysis of lived experience, i.e. of subjective phenomena, has to be an integral part of any science of consciousness.[13] This amounts to a profound change of methodology, which many cognitive scientists are reluctant to embrace, and which lies at the very root of the "hard problem of consciousness."

The great reluctance of scientists to deal with subjective phenomena is part of our Cartesian heritage. Descartes's fundamental division between mind and matter, between the I and the world, made us believe that the world could be described objectively, i.e. without ever mentioning the human observer. Such an objective description of nature became the ideal of all science. However, three centuries after Descartes, quantum theory showed us that this classical ideal of an objective science cannot be maintained when dealing with atomic phenomena. And more recently, the Santiago Theory of Cognition has made it clear that cognition itself is not a representation of an independently existing world, but rather a "bringing forth" of a world through the process of living.

We have come to realize that the subjective dimension is always implicit in the practice of science, but in general it is not the explicit focus. In a science of consciousness, by contrast, some of the very data to be examined are subjective, inner experiences. To collect and analyze these data systematically requires a disciplined examination of first-person subjective experience. Only when such an examination becomes an integral part of the study of consciousness will it deserve to be called a "science of consciousness."

This does not mean that we have to give up scientific rigor. When we speak of an "objective description" in science, we mean first and foremost a body of knowledge that is shaped, constrained, and regulated by collective scientific enterprise, rather than merely a collection of individual accounts. Even when the object of investigation consists of first-person accounts of conscious experience, the intersubjective validation that is standard practice in science need not be abandoned.[14]

Schools of Consciousness Study

The use of complexity theory and the systematic analysis of first-person conscious experience will be crucial in formulating a proper science of consciousness. In the last few years, several significant steps have already been taken toward this goal. Indeed, the extent to which nonlinear dynamics and the analysis of first-person experience are utilized can be used to identify several broad schools of thought among the great variety of current approaches to the study of consciousness.[15]

The first is the most traditional school of thought. It includes, among others, the neuroscientist Patricia Churchland and the molecular biologist and Nobel laureate Francis Crick.[16] This school has been called "neuroreductionist" by Francisco Varela, because it reduces consciousness to neural mechanisms. Thus, consciousness is "explained away," as Churchland puts it, much like heat in physics was explained away once it was recognized as the energy of molecules in motion. In the words of Francis Crick:

> "You," your joys and your sorrows, your memories and your ambitions, your sense of personal identity and free will, are in fact no

> more than the behavior of a vast assembly of nerve cells and their associated molecules. As Lewis Carroll's Alice might have phrased it: "You're nothing but a pack of neurons."[17]

Crick explains in detail how consciousness is reduced to the firing of neurons, and he also asserts that conscious experience is an emergent property of the brain as a whole, but he never addresses the nonlinear dynamics of this process of emergence, and thus remains unable to solve the "hard problem of consciousness." As philosopher John Searle formulates the challenge, "How is it possible for physical, objective, quantitatively describable neuron firings to cause qualitative, private, subjective experiences?"[18]

The second school of consciousness study, known as "functionalism," is the most popular among today's cognitive scientists and philosophers.[19] Its proponents assert that mental states are defined by their "functional organization," i.e. by patterns of causal relations in the nervous system. The functionalists are not Cartesian reductionists, because they pay careful attention to nonlinear neural patterns, but they deny that conscious experience is an irreducible, emergent phenomenon. It may seem an irreducible experience, but in their view a conscious state is defined completely by its functional organization and is therefore understood once that pattern of organization has been identified. Daniel Dennett, one of the leading functionalists, gave his book the catchy title *Consciousness Explained*.[20]

Many patterns of functional organization have been postulated by cognitive scientists, and consequently there are many different strands of functionalism today. Sometimes analogies between functional organization and computer software, derived from artificial intelligence research, are also included among the functionalist approaches.[21]

Less known is a small school of philosophers who call themselves "mysterians." They argue that consciousness is a deep mystery which human intelligence, because of its inherent limitations, will never unravel.[22] At the root of these limitations, in their view, lies an irreducible duality, which turns out to be the classical Cartesian duality between mind and matter. While introspection cannot teach us anything about the brain as a physical object, the study of brain structure cannot give us any access to conscious experience. Because they neglect to view consciousness as a process and do not appreciate the nature of an emergent phenomenon, the mysterians are unable to bridge the Cartesian gap and conclude that the nature of consciousness will forever remain a mystery.

Finally, there is a small but growing school of consciousness studies that embraces both the use of complexity theory and the analysis of first-person experience. Francisco Varela, one of the leaders of this school of thought, has given it the name "neurophenomenology."[23] Phenomenology is an important branch of modern philosophy, founded by Edmund Husserl at the beginning of the twentieth century and developed further by many European philosophers, including Martin Heidegger and Maurice Merleau-Ponty. The central concern of phenomenology is the disciplined examination of experience, and the hope of Husserl and his followers was, and is, that a true science of experience would eventually be established in partnership with the natural sciences.

Neurophenomenology is an approach to the study of consciousness that combines the disciplined examination of conscious experience with the analysis of corresponding neural patterns and processes. With this dual approach, neurophenomenologists explore various domains of experience and try to understand how they emerge from complex neural activities. In doing so, these cognitive scientists are taking the first steps toward formulating a true science of experience. It has been very gratifying for me personally to realize that their project has much in common with the science of consciousness I envisaged more than twenty years ago in a conversation with the psychiatrist R. D. Laing, when I speculated that

> a true science of consciousness . . . would have to be a new type of science dealing with qualities rather than quantities and being based on shared experience rather than verifiable measurements. The data of such a science would be patterns of experience that cannot be quantified or analysed. On the other hand, the conceptual models in-

terconnecting the data would have to be logically consistent, like all scientific models, and might even include quantitative elements.[24]

The View from Within

The basic premise of neurophenomenology is that brain physiology and conscious experience should be treated as two interdependent domains of research with equal status. The disciplined examination of experience and the analysis of the corresponding neural patterns and processes will generate reciprocal constraints, so that research activities in the two domains can guide one another in a systematic exploration of consciousness.

Today's neurophenomenologists are a very diverse group. They differ in the manner in which first-person experience is taken into account, and they have also proposed different models for the corresponding neural processes. The whole field is presented in some detail in a special issue of the *Journal of Consciousness Studies*, titled "The View From Within" and edited by Francisco Varela and Jonathan Shear.[25]

As far as first-person experience is concerned, three main approaches are being pursued. The first is introspection, a method developed at the very beginning of scientific psychology. The second is the phenomenological approach in the strict sense, as developed by Husserl and his followers. The third approach consists in using the wealth of evidence gathered from meditative practice, especially within the Buddhist tradition. Whatever their approach, these cognitive scientists insist that they are not talking about a casual inspection of experience, but about using strict methodologies that require special skills and sustained training, just like the methodologies in other areas of scientific observation.

The methodology of introspection was advocated as the primary tool of psychology by William James at the end of the nineteenth century and was standardized and practiced with great enthusiasm during the subsequent decades. However, it soon ran into difficulties—not because of any intrinsic flaws, but because the data it produced were in strong disagreement with the hypotheses formulated at the outset.[26] The observations were far ahead of the theoretical ideas of the time, and rather than reexamine their theories, psychologists criticized each others' methodologies, which led to a general distrust of the whole practice of introspection. As a result, half a century passed without any developments or improvements in introspective practice.

Today, the methods developed by the pioneers of introspection are mostly found in the practices of psychotherapists and professional trainers without any connections to academic research programs in cognitive science. A small group of cognitive scientists is now attempting to revive this dormant tradition for a systematic and sustained exploration of conscious experience.[27]

By contrast, phenomenology was developed by Edmund Husserl as a philosophical discipline rather than a scientific method. Its central characteristic is a specific gesture of reflection, known as "phenomenological reduction."[28] This term must not be confused with reductionism in the natural sciences. In the philosophical sense, reduction (from the Latin *reducere*) means a "leading back," or disengaging of subjective experience, through the suspension of beliefs about what is being experienced. In this way, the field of experience appears more vividly present and a capacity for systematic reflection is cultivated. In philosophy, this is known as the shift from the natural to the phenomenological attitude.

To anybody with some experience in meditation practice, this description of the phenomenological attitude will have a familiar ring. Indeed, contemplative traditions have developed rigorous techniques for examining and probing the mind for centuries, and have shown that these skills can be refined considerably over time. Throughout human history, the disciplined examination of experience has been used within widely differing philosophical and religious traditions, including Hinduism, Buddhism, Taoism, Sufism, and Christianity. We may therefore expect that some of the insights of these traditions will be valid beyond their particular metaphysical and cultural frameworks.[29]

This applies especially to Buddhism, which has flourished in many different cultures, originating with the Buddha in India, then spreading to China and Southeast Asia, ending up in Japan, and, many centuries later, crossing the Pacific to California. In these different cultural contexts, mind and consciousness have always been the primary objects of Buddhist contemplative investigations. Buddhists regard the undisciplined mind as an unreliable instrument for observing different states of consciousness, and, following the Buddha's initial instructions, they have developed a great variety of techniques for stabilizing and refining the attention.[30]

Over the centuries, Buddhist scholars have formulated elaborate and sophisticated theories about many subtle aspects of conscious experience, which are likely to be fertile sources of inspiration for cognitive scientists. The dialogue between cognitive science and Buddhist contemplative traditions has already begun, and the first results indicate that evidence from meditative practices will be a valuable component of any future science of consciousness.[31]

The schools of consciousness study mentioned above all share the basic insight that consciousness is a cognitive process, emerging from complex neural activity. However, there are also other attempts, mostly by physicists and mathematicians, to explain consciousness as a direct property of matter, rather than as a phenomenon associated with life. An outstanding example of that position is the approach of the mathematician and cosmologist Roger Penrose, who postulates that consciousness is a quantum phenomenon and claims that "we don't understand consciousness, because we don't understand enough about the physical world."[32]

These views of "mind without biology," in the apt phrase of neuroscientist and Nobel laureate Gerald Edelman,[33] also include the view of the brain as a complicated computer. Like many cognitive scientists, I believe that these are extreme views that are fundamentally flawed and that conscious experience is an expression of life, emerging from complex neural activity.[34]

Consciousness and the Brain

Let us now turn to the neural activity that underlies conscious experience. In recent years, cognitive scientists have made significant advances in identifying the links between neurophysiology and the emergence of experience. In my opinion, the most promising models have been proposed by Francisco Varela and, more recently, by Gerald Edelman in collaboration with Giulio Tononi.[35]

In both cases, the authors cautiously present their models as hypotheses, and the core idea of the two hypotheses is the same. Conscious experience is not located in a specific part of the brain, nor can it be identified in terms of special neural structures. It is an emergent property of a particular cognitive process—the formation of transient functional clusters of neurons. Varela calls such clusters "resonant cell assemblies," while Tononi and Edelman speak of a "dynamic core."

It is also interesting to notice that Tononi and Edelman embrace the basic premise of neurophenomenology that brain physiology and conscious experience should be treated as two interdependent domains of research. "It is a central claim of this article," they write, "that analyzing the convergence between . . . phenomenological and neural properties can yield valuable insights into the kinds of neural processes that can account for the corresponding properties of conscious experience."[36]

The detailed dynamics of the neural processes in these two models are different but, perhaps, not incompatible. They differ in part because the authors do not focus on the same characteristics of conscious experience, and hence emphasize different properties of the corresponding neural clusters.

Varela starts from the observation that the "mental space" of a conscious experience is composed of many dimensions. In other words, it is created by many different brain functions, and yet is a single coherent experience. For example, when the smell of a perfume evokes a pleasant or unpleasant sensation, we experience this conscious state as an integrated whole, composed of sensory perceptions, memories and emotions. The experience is not constant, as we well know, and may be extremely short. Conscious states are transitory, continually arising and subsiding. Another important observation is that the experiential state is always "embodied," that is, embedded in a particular field of sensation. In fact, most conscious states seem to have a dominant sensation that colors the entire experience.[37]

The specific neural mechanism proposed by Varela for the emergence of transitory experiential states is a resonance phenomenon known as "phase-locking," in which different brain regions are interconnected in such a way that their neurons fire in synchrony. Through this synchronization of neural activity, temporary "cell assemblies" are formed, which may consist of widely dispersed neural circuits.

According to Varela's hypothesis, each conscious experience is based on a specific cell assembly, in which many different neural activities—associated with sensory perception, emotions, memory, bodily movements, etc.—are unified into a transient but coherent ensemble of oscillating neurons. The best way to think of this neural process is, perhaps, in musical terms.[38] There are noises; then they come together in synchrony as a melody emerges; then the melody subsides again into cacophony, until another melody arises in the next moment of resonance.

Varela has applied his model in considerable detail to the exploration of the experience of present time—a traditional theme in phenomenological studies—and has suggested similar explorations of other aspects of conscious experience.[39] These include various forms of attention and the corresponding neural networks and pathways; the nature of will, as expressed in the initiation of voluntary action; and the neural correlates of emotions, as well as the relationships between mood, emotion, and reason. According to Varela, progress in such a research program will depend largely on the extent to which cognitive scientists are willing to build a sustained tradition of phenomenological examination.

Let us now turn to the neural processes described in the model by Gerald Edelman and Giulio Tononi. Like Francisco Varela, these authors emphasize that conscious experience is highly integrated, each conscious state comprising a single "scene" that cannot be decomposed into independent components. In addition, they point out that conscious experience is also highly differentiated, in the sense that we can experience any of a huge number of different conscious states within a short time. These observations provide two criteria for the underlying neural processes: they have to be integrated, while showing extraordinary differentiation, or complexity.[40]

The mechanism the authors propose for the rapid integration of neural processes in different areas of the brain is one that has been developed theoretically by Edelman since the 1980s and has been tested extensively in large-scale computer simulations by Edelman, Tononi, and their colleagues. It is called "reentry" and consists of continual exchanges of parallel signals within and among brain areas.[41] These processes of parallel signaling play the same role as the phase-locking in Varela's model. Indeed, as Varela speaks of cell assemblies being "glued together" by phase-locking, so Tononi and Edelman speak of a dynamic "binding" of groups of nerve cells through the process of reentry.

Conscious experience emerges, according to Tononi and Edelman, when the activities of different brain areas are integrated during brief moments through the process of reentry. Each conscious experience emerges from a functional cluster of neurons, which together constitute a unified neural process, or "dynamic core." The authors chose the term "dynamic core" to convey both the idea of integration and of constantly changing activity patterns. They emphasize that the dynamic core is not a thing or a location, but a process of varying neural interactions.

A dynamic core may change its composition over time, and the same group of neurons may at times be part of a dynamic core and thus underlie conscious experience, and at other times not be part of it and thus be involved in unconscious processes. Moreover, since the core is a cluster of neurons that are functionally integrated without necessarily being adjacent anatomically, the composition of the core can transcend traditional anatomic boundaries. Finally, the exact composition of the dynamic core associated with a particular conscious experience is expected to vary from individual to individual.

In spite of the differences in the detailed dynamics they describe, the two hypotheses of resonant cell assemblies and the dynamic core evidently have much in common. Both view conscious experience as an emergent property of a transient process of integration, or synchronization, of widely distributed groups of neurons. Both offer concrete, testable proposals for the specific dynamics of that process, and are likely to lead to significant advances in the formulation of a proper science of consciousness in the years to come.

The Social Dimension of Consciousness

As human beings, we not only experience the integrated states of primary consciousness; we also think and reflect, communicate through symbolic language, make value judgments, hold beliefs, and act intentionally with self-awareness and an experience of personal freedom. Any future theory of consciousness will have to explain how these well-known characteristics of the human mind arise out of the cognitive processes that are common to all living organisms.

As I mentioned above, the "inner world" of our reflective consciousness emerged in evolution together with language and social reality.[42] This means that human consciousness is not only a biological, but also a social, phenomenon. The social dimension of reflective consciousness is frequently ignored by scientists and philosophers. As cognitive scientist Rafael Núñez points out, almost all current views of cognition implicitly assume that the appropriate unit of analysis is the body and the mind of the individual.[43] This tendency has been reinforced by the new technologies for analyzing brain functions, which invite cognitive scientists to study single, isolated brains and to neglect the continual interactions of those brains with other bodies and brains within communities of organisms. These interactive processes are crucial to understanding the level of cognitive abstraction that is characteristic of reflective consciousness.

Humberto Maturana was one of the first scientists to link the biology of human consciousness to language in a systematic way.[44] He did so by approaching language through a careful analysis of communication within the framework of the Santiago Theory of Cognition. Communication, according to Maturana, is not the transmission of information but rather the coordination of behavior between living organisms through mutual structural coupling.[45] In these recurrent interactions, the living organisms change together through their mutual triggering of structural changes. Such mutual coordination is the key characteristic of communication for all living organisms, with or without nervous systems, and it becomes more and more subtle and elaborate with nervous systems of increasing complexity.

Language arises when a level of abstraction is reached at which there is communication about communication. In other words, there is a coordination of coordinations of behavior. For example (as Maturana explained in a seminar), when you hail a taxi driver on the other side of the street with a gesture of your hand, thereby getting his attention, this is a coordination of behavior. When you then describe a circle with your hand, asking him to make a U-turn, this coordinates the coordination, and thus arises the first level of communication in language. The circle has become a symbol, representing your mental image of the taxi's trajectory. This little example illustrates the important point that language is a system of symbolic communication. Its symbols—words, gestures, and other signs—serve as tokens for the linguistic coordination of actions. This, in turn, creates the notion of objects, and thus the symbols become associated with our mental images of objects.

Then, as soon as words and objects are created through coordinations of coordinations of behavior, they become the basis for further coordinations, which generate a series of recursive levels of linguistic communication.[46] As we distinguish objects, we create abstract concepts to denote their properties, as well as the relations between objects. The process of observation, according to Maturana, consists of such distinctions of distinctions; then the observer appears when we distinguish between observations; and, finally, self-awareness arises as the observation of the observer, when we use the notion of an object and the associated abstract concepts to describe ourselves. Thus our linguistic domain expands to include reflective consciousness. At each of these recursive levels words and objects are generated, and their distinction then obscures the coordinations which they coordinate.

Maturana emphasizes that the phenomenon of language does not occur in the brain, but in a continual flow of coordinations of coordinations of behavior. It occurs, in Maturana's words, "in the flow of interactions and relations of living together."[47] As humans, we exist in language and we continually weave the linguistic web in which we are embedded. We coordinate our behavior in language, and together in language we bring forth our world. "The world everyone sees," write Maturana and Varela, "is not *the* world but *a* world, which we bring forth with others."[48] This human world centrally includes our inner world of abstract thought, concepts, beliefs, mental images, intentions, and self-awareness. In a human conversation, our concepts and ideas, emotions and body movements become tightly linked in a complex choreography of behavioral coordination.

Conversations with Chimpanzees

Maturana's theory of consciousness establishes a set of crucial links between self-awareness, conceptual thought and symbolic language. On the basis of this theory, and in the spirit of neurophenomenology, we can now ask: What is the neurophysiology underlying the emergence of human language? How did we, in our human evolution, develop the extraordinary levels of abstraction that are characteristic of our thought and language? The answers to these questions are still far from definite, but several dramatic insights have emerged over the last two decades that force us to revise many long-cherished scientific and philosophical assumptions.

One radically new way of thinking about human language is suggested by several decades of research into communication with chimpanzees through sign language. Psychologist Roger Fouts, one of the pioneers at the very center of this research, published a fascinating account of his groundbreaking work in his book *Next of Kin.*[49] Fouts not only tells the enthralling story of how he personally experienced extended dialogues between humans and apes, but also uses the insights he gained to offer some very exciting speculations about the evolutionary origins of human language.

Recent DNA research has shown that there is only a 1.6 percent difference between human DNA and chimpanzee DNA. Indeed, chimpanzees are more closely related to humans than to gorillas or orangutans. As Fouts explains: "Our skeleton is an upright version of the chimpanzee skeleton; our brain is an enlarged version of the chimpanzee brain; our vocal tract is an innovation on the chimpanzee vocal tract."[50] In addition, it is well known that much of the chimps' facial repertoire is similar to our own.

The DNA evidence we have today strongly indicates that chimpanzees and humans share a common ancestor which the gorillas do not share. If we classify the chimpanzees as great apes, then we must classify ourselves as great apes, too. Indeed, any category of ape is meaningless unless it includes humans. The Smithsonian Institute has changed its classification scheme accordingly. In the most recent edition of its publication *Mammal Species of the World*, the members of the great ape family have been moved into the family of hominids, which was previously reserved for humans alone.[51]

The continuity between humans and chimpanzees does not end with anatomy, but also extends to social and cultural characteristics. Like us, chimpanzees are social creatures. In captivity, they suffer most from loneliness and boredom. In the wild, they thrive on change, foraging in different fruit trees every day, building different sleeping nests every night, and socializing with various members of their community as they travel through the jungle.

Moreover, anthropologists have been amazed to discover that chimpanzees also have distinct cultures. Since Jane Goodall made the momentous discovery in the late 1950s that wild chimpanzees make and use tools, extensive observations have revealed that chimpanzee communities have unique hunter-gatherer cultures, in which the young learn new skills from their mothers through a combination of imitation and guidance.[52] Some of the hammers and anvils they use to crack nuts are identical to the tools of our own hominid ancestors, and the style of toolmaking differs from community to community, as it did in the early hominid communities.

Anthropologists have also documented the chimpanzees' widespread use of medicinal plants, and some scientists believe that there may be dozens of local chimp medicine cultures scattered across Africa. In addition, chimpanzees nurture family bonds, mourn the death of mothers and adopt orphans, struggle for power and wage war. In short, there seems to be as much social and cultural continuity in the evolution of humans and chimpanzees as there is anatomical continuity.

So, what about cognition and language? For a long time, scientists assumed that chimpanzee communication had nothing to do with human communication because the chimps' grunts and screams bear little resemblance to human speech. However, as Roger Fouts argues eloquently, these scientists focused on the wrong channel of communication.[53] Careful observation of chimpanzees in the wild has shown that they use their hands for much more than building tools. They are communicating with them in ways previously unimagined, gesturing to beg for food, to seek reassurance, and to offer encouragement. There are various chimpanzee gestures for "Come with me," "May I pass?" and "You are welcome," and, most astonishingly, some of these gestures differ from community to community.

These observations were dramatically confirmed by the results of several teams of psychologists who spent many years raising chimpanzees in their homes like human children, while communicating with them in American Sign Language (ASL). Fouts emphasizes that, to appreciate the implications of this research, it is important to understand that ASL is not an artificial system that hearing people invented for the deaf. It has existed for at least 150 years and has its roots in various European sign languages that were developed by the deaf themselves over centuries.

Like spoken languages, ASL is highly flexible. Its building blocks—hand configurations, placements, and movements—can be combined to form an infinite number of signs, the equivalent of words. ASL has its own rules for organizing signs into sentences, exhibiting a subtle and complex visual grammar that is very different from English grammar.[54]

In the cross-fostering studies with chimpanzees, young chimps were not treated as passive laboratory subjects, but as primates endowed with a powerful need to learn and communicate. It was hoped that they would not only acquire a rudimentary ASL vocabulary and grammar, but would also use it to ask questions, comment on their experiences, and stimulate conversations. In other words, the scientists aimed to engage in a genuine two-way communication with the apes. And this is what happened.

Roger Fouts's first and most famous "foster child" was a young chimp called Washoe, who at the age of four was able to use ASL at the level of a two- or three-year-old human child. Like any human toddler, Washoe often greeted her "parents" with a flurry of messages—ROGER HURRY, COME HUG, FEED ME, GIMME CLOTHES, PLEASE OUT, OPEN DOOR—and like all small children, she also talked to her pets and her dolls, and even to herself. For Fouts, "Washoe's spontaneous 'hand chatter' was the most compelling evidence that she was using language the way human children do . . . The way [she] ran on with her hands like a gregarious deaf child, sometimes in the most unlikely of circumstances, caused more than one sceptic to reconsider his long-cherished assumption that animals can neither think nor talk."[55]

When Washoe grew into an adult ape, she taught her adopted son how to sign, and later on, when they both lived together with three other chimpanzees of various ages, they formed a complex and cohesive family in which language flourished quite naturally. Roger Fouts and his wife and collaborator, Deborah Harris Fouts, randomly videotaped many hours of animated chimpanzee conversations. These tapes show Washoe's family signing while they share blankets, play games, eat breakfast, and get ready for bed. As Fouts tells it, "The chimps were signing to one another even in the middle of screaming family fights, which was the best indication that sign language had become an integral part of their mental and emotional lives." Fouts also reports that the chimps' conversations were so clear that independent ASL experts agreed nine out of ten times on the meanings of these videotaped exchanges.[56]

The Origins of Human Language

The unprecedented dialogues between humans and chimpanzees opened a unique window into the apes' cognitive abilities that sheds new light on the origins of human language. As Fouts documents in great detail, his work with chimpanzees over several decades has shown that they can use abstract symbols and metaphors, have a mental grasp of classifications and understand simple grammar. They are also able to use syntax, i.e. to combine symbols in an order that conveys meaning, and they creatively combine signs in new ways to invent new words.

These stunning discoveries led Roger Fouts to revive a theory of the origin of human language advanced by anthropologist Gordon Hewes in the early 1970s.[57] Hewes proposed that early hominids communicated with their hands and developed the skill of precise hand movements both for gestures and for making tools. Speech would have evolved later from the capacity for "syntax"—an ability to follow complex patterned sequences in the making of tools, in gesturing and in forming words.

These insights have very interesting implications for the understanding of technology. If language originated in gesture, and if gesture and toolmaking (the simplest form of technology) evolved together, this would imply that technology is an essential part of human nature, inseparable from the evolution of language and consciousness. It would mean that, from the very dawn of our species, human nature and technology have been inseparably linked.

The idea that language may have originated in gesture is, of course, not new. For centuries people have noticed that infants begin gesturing before they begin speaking, and that gesture is a universal means of communication we can always fall back on when we do not speak the same language. The scientific problem was to understand how speech could have evolved physically out of gestures. How did our hominid ancestors bridge the gap between motions of the hand and streams of words from the mouth?

This puzzle was solved by neurologist Doreen Kimura, when she discovered that speech and precise hand movements seem to be controlled by the same motor region of the brain.[58] When Fouts learned about Kimura's discovery, he realized that, in a sense, sign language and spoken language are both forms of gesture. In his words: "Sign language uses gesture of the hands; spoken language is gesture of the tongue. The tongue makes precise movements, stopping at specific places around the mouth so that we can produce certain sounds. The hands and fingers stop at precise places around the body to produce signs."[59]

The realization enabled Fouts to formulate his basic theory of the evolutionary origin of spoken language. Our hominid ancestors must have communicated with their hands, just as their ape cousins did. Once they began to walk upright, their hands were free to develop more elaborate and refined gestures. Over time, their gestural grammar would have become more and more complex, as the gestures themselves evolved from gross to more precise movements. Eventually, the precise movements of their hands would have triggered precise movements of their tongues, and thus the evolution of gesture produced two important dividends: the ability to make and use more complex tools, and the ability to produce sophisticated vocal sounds.[60]

This theory was confirmed dramatically when Roger Fouts began to work with autistic children.[61] His work with chimpanzees and sign language had made him realize that, when doctors say that autistic children have "language problems," they really mean that these children have problems with *spoken* language. So, Fouts introduced sign language as an alternative linguistic channel, just as he had done with the chimps. He had extraordinary success with this technique. After a couple of months of signing, the children broke through their isolation and their behavior changed dramatically.

Even more extraordinary, and at first totally unexpected, was the fact that the autistic children began to speak after several weeks of signing. The signing apparently triggered the capacity for speech. The skill of forming precise signs could be transferred to the skill of forming sounds because both are controlled by the same brain structures. "In a matter of weeks," Fouts concluded, the children "may very well have retraced the evolutionary path of our own ancestors, a six-million-year journey that led from apelike gesture to modern human speech."[62]

Fouts speculates that humans began shifting to speech about 200,000 years ago with the evolution of the so-called "archaic forms" of *homo sapiens.* That date coincides with the first fabrications of specialized stone tools that required considerable manual dexterity. The early humans who produced these tools were likely to possess the kind of neural mechanisms that would have also enabled them to produce words.

The appearance of vocal words in our ancestors' communication brought immediate advantages. Those who communicated vocally could do so when their hands were full, or when the listener's back was turned. Eventually, those evolutionary advantages would bring about the anatomical changes that were necessary for full-blown speech. Over tens of thousands of years, as our vocal tracts evolved, humans communicated through combinations of precise gestures and spoken words until, eventually, the spoken words crowded out the signs and became the dominant form of human communication. Even today, however, we use gestures whenever spoken language does not serve us. "As our species' oldest form of communication," Fouts observes, "gesture still functions as every culture's 'second language.' "[63]

The Embodied Mind

According to Roger Fouts, then, language was originally embodied in gesture and evolved from gesture together with human consciousness. This theory is consistent with the recent discovery by cognitive scientists that conceptual thought as a whole is embodied physically in the body and brain.

When cognitive scientists say that the mind is embodied, they mean far more than the obvious fact that we need a brain in order to think. Recent studies in the new field of cognitive linguistics indicate strongly that human reason does not transcend the body, as much of Western philosophy has held, but is shaped crucially by our physical nature and our bodily experience. It is in that sense that the human mind is fundamentally embodied. The very structure of reason arises from our bodies and brains.[64]

The evidence for the mind's embodiment and the profound philosophical implications of this insight are presented lucidly and eloquently by two leading cognitive linguists, George Lakoff and Mark Johnson, in their book *Philosophy in the Flesh*.[65] The evidence is based, first of all, on the discovery that most of our thought is unconscious, operating at a level that is inaccessible to ordinary conscious awareness. This "cognitive unconscious" includes not only all our automatic cognitive operations, but also our tacit knowledge and beliefs. Without our awareness, the cognitive unconscious shapes and structures all conscious thought. This has become a major field of study in cognitive science, which has resulted in radically new views of how concepts and thought processes are formed.

At this point, the detailed neurophysiology of the formation of abstract concepts is still unclear. However, cognitive scientists have begun to understand one crucial aspect of this process. In the words of Lakoff and Johnson: "The same neural and cognitive mechanisms that allow us to perceive and move around also create our conceptual structures and modes of reason."[66]

This new understanding of human thought began in the 1980s with several studies of the nature of conceptual categories.[67] The process of categorizing a variety of experiences is a fundamental part of cognition at all levels of life. Microorganisms categorize chemicals into food and nonfood, into what to move toward and what to move away from. Similarly, animals categorize food, noises that mean danger, members of their own species, sexual signals, and so on. As Maturana and Varela would say, a living organism brings forth a world by making distinctions.

How living organisms categorize depends on their sensory apparatus and their motor systems; in other words, it depends on how they are embodied. This is true not only for animals, plants, and microorganisms, but also for human beings, as cognitive scientists have recently discovered. Although some of our categories are the result of conscious reasoning, most of them are formed automatically and unconsciously as a result of the specific nature of our bodies and brains.

This can easily be illustrated with the example of colors. Extensive studies of color perception over several decades have made it clear that there are no colors in the external world, independent of the process of perception. Our experience of color is created by the wave-lengths of reflected light in interaction with the color cones in our retinas and the neural circuitry connected to them. Indeed, detailed studies have shown that the entire structure of our color categories (the number of colors, hues, etc.) arises from our neural structures.[68]

Whereas color categories are based on our neurophysiology, other types of categories are formed on the basis of our bodily experience. This is especially important for spatial relations, which are among our most basic categories. As Lakoff and Johnson explain, when we perceive a cat "in front of" a tree, this spatial relationship does not exist objectively in the world, but is a projection from our bodily experience. We have bodies with inherent fronts and backs, and we project this distinction onto other objects. Thus, "our bodies define a set of fundamental spatial relations that we use not only in orienting ourselves, but in perceiving the relationship of one object to another."[69]

As human beings, we not only categorize the varieties of our experience, but also use abstract concepts to characterize our categories and reason about them. At the human level of cognition, categories are always conceptual—inseparable from the corresponding abstract concepts. And since our categories arise from our neural structures and bodily experience, so do our abstract concepts.

Some of our embodied concepts are also the basis of certain forms of reasoning, which means that the way we think is also embodied. For example, when we distinguish between "inside" and "outside," we tend to visualize this spatial relationship in terms of a container with an inside, a boundary, and an outside. This mental image, which is grounded in the experience of our body as a container, becomes the basis of a certain form of reasoning.[70] Suppose we put a cup inside a bowl and a cherry inside the cup. We would know immediately, just by looking at it, that the cherry, being inside the cup, is also inside the bowl.

That inference corresponds to a well-known argument, or "syllogism," in classical Aristotelian logic. In its most familiar form, it goes: "All men are mortal. Socrates is a man. Therefore, Socrates is mortal." The argument seems conclusive because, like our cherry, Socrates is within the "container" (category) of men, and men are within the "container" (category) of mortals. We project the mental image of containers onto abstract categories, and then use our bodily experience of a container to reason about these categories.

In other words, the classical Aristotelian syllogism is not a form of disembodied reasoning, but grows out of our bodily experience. Lakoff and Johnson argue that this is true for many other forms of reasoning as well. The structures of our bodies and brains determine the concepts we can form and the reasoning we can engage in.

When we project the mental image of a container onto the abstract concept of a category, we use it as a metaphor. This process of metaphorical projection is a crucial element in the formation of abstract thought and the discovery that most human thought is metaphorical has been another major advance in cognitive science.[71] Metaphors make it possible to extend our basic embodied concepts into abstract theoretical domains. When we say, "I don't seem to be able to grasp this idea," or "This is way over my head," we use our bodily experience of grasping an object to reason about understanding an idea. In the same way, we speak of a "warm welcome" or a "big day," projecting sensory and bodily experiences onto abstract domains.

These are all examples of primary metaphors—the basic elements of metaphorical thought. Cognitive linguists theorize that we acquire most of our primary metaphors automatically and unconsciously in our early childhood.[72] For infants, the experience of affection typically occurs together with that of warmth, of being held. Thus associations between the two experiential domains are built up, and corresponding pathways across neural networks are established. Later in life, these associations continue as metaphors when we speak of a "warm smile" or a "close friend."

Our thought and language contain hundreds of primary metaphors, most of which we use without ever being aware of them; and since they originate in basic bodily experiences, they tend to be the same in most languages around the world. In our abstract thought processes, we combine primary metaphors into more complex ones, which enables us to use rich imagery and subtle conceptual structures when we reflect on our experience. For example, to think of life as a journey allows us to use our rich knowledge of journeys while reflecting on how to lead a purposeful life.[73]

Human Nature

During the last two decades of the twentieth century, cognitive scientists made three major discoveries. As Lakoff and Johnson summarize: "The mind is inherently embodied. Thought is mostly unconscious. Abstract concepts are largely metaphorical."[74] When these insights are widely accepted and integrated into a coherent theory of human cognition, they will force us to reexamine many of the principal tenets of Western philosophy. In *Philosophy in the Flesh* the authors take the first steps toward such a rethinking of Western philosophy in the light of cognitive science.

Their main argument is that philosophy should be able to respond to the fundamental human need to know ourselves—to know "who we are, how we experience the world, and how we ought to live." Knowing ourselves includes understanding how we think and how we express our thoughts in language, and it is here that cognitive science can make important contributions to philosophy. "Since everything we think and say and do depends on the workings of our embodied minds," Lakoff and Johnson argue, "cognitive science is one of our most profound resources for self-knowledge."[75]

The authors envisage a dialogue between philosophy and cognitive science in which the two disciplines support and enrich each other. Scientists need philosophy to become aware of how hidden philosophical assumptions influence their theories. As John Searle reminds us, "The price of having contempt for philosophy is that you make philosophical mistakes."[76] Philosophers, on the other hand, cannot propose serious theories about the nature of language, mind, and consciousness without taking into account the recent remarkable advances in the scientific understanding of human cognition.

In my view, the main significance of these advances has been the gradual but consistent healing of the Cartesian split between mind and matter that has plagued Western science and philosophy for more than 300 years. The Santiago Theory has shown that at all levels of life, mind and matter, process and structure, are inseparably connected.

Recent research in cognitive science has confirmed and refined this view by showing how the process of cognition evolved into forms of increasing complexity together with the corresponding biological structures. As the ability to control precise hand and tongue movements developed, language, reflective consciousness, and conceptual thought evolved in the early humans as parts of ever more complex processes of communication.

All these are manifestations of the process of cognition, and at each new level they involve corresponding neural and bodily structures. As the recent discoveries in cognitive linguistics have shown, the human mind, even in its most abstract manifestations, is not separate from the body but arises from it and is shaped by it.

The unified, post-Cartesian view of mind, matter, and life also implies a radical reassessment of the relationship between humans and animals. Throughout most of Western philosophy, the capacity to reason was seen as a uniquely human characteristic, distinguishing us from all other animals. The communication studies with chimpanzees have exposed the fallacy of this belief in the most dramatic of ways. They make it clear that the cognitive and emotional lives of animals and humans differ only by degree; that life is a great continuum in which differences between species are gradual and evolutionary. Cognitive linguists have fully confirmed this evolutionary conception of human nature. In the words of Lakoff and Johnson, "Reason, even in its most abstract form, makes use of, rather than transcends, our animal nature. Reason is thus not an essence that separates us from other animals; rather, it places us on a continuum with them."[77]

The Spiritual Dimension

The scenario of the evolution of life that I discussed in the preceding pages begins with the formation of membrane-bounded bubbles in the primeval oceans. These tiny droplets formed spontaneously in an appropriate soap-and-water environment, following the basic laws of physics and chemistry. Once they had formed, a complex network chemistry gradually unfolded in the spaces they enclosed, which provided the bubbles with the potential to grow and evolve into complex, self-replicating structures. When catalysts entered the system, molecular complexity increased rapidly, and eventually life emerged from these protocells with the evolution of proteins, nucleic acids, and the genetic code.

This marked the emergence of a universal ancestor—the first bacterial cell—from which all subsequent life on Earth descended. The descendants of the first living cells took over the Earth by weaving a planetary bacterial web and gradually occupying all the ecological niches. Driven by the creativity inherent in all living systems, the planetary web of life expanded through mutations, gene trading, and symbioses, producing forms of life of ever-increasing complexity and diversity.

In this majestic unfolding of life, all living organisms continually responded to environmental influences with structural changes, and they did so autonomously, according to their own natures. From the beginning of life, their interactions with one another and with the nonliving environment were cognitive interactions. As their structures increased in complexity, so did their cognitive processes, eventually bringing forth conscious awareness, language, and conceptual thought.

When we look at this scenario—from the formation of oily droplets to the emergence of consciousness—it may seem that all there is to life is molecules, and the question naturally arises: What about the spiritual dimension of life? Is there any room in this new vision for the human spirit?

The view that life, ultimately, is all about molecules is one that is often advanced by molecular biologists. It is important to realize, in my opinion, that this is a dangerously reductionist view. The new understanding of life is a systemic understanding, which means that it is based not only on the analysis of molecular structures, but also on the analysis of patterns of relationships among these structures and of the specific processes underlying their formation. As we have seen, the defining characteristic of a living system is not the presence of certain macromolecules, but the presence of a self-generating network of metabolic processes.[78]

The processes of life include, most importantly, the spontaneous emergence of new order, which is the basis of life's inherent creativity. Moreover, the life processes are associated with the cognitive dimension of life, and the emergence of new order includes the emergence of language and consciousness.

Where does the human spirit come into this picture? To answer this question, it will be useful to review the original meaning of "spirit." As we have seen, the Latin *spiritus* means "breath," which is also true for the related Latin word *anima*, the Greek *psyche*, and the Sanskrit *atman*.[79] The common meaning of these key terms indicates that the original meaning of spirit in many ancient philosophical and religious traditions, in the West as well as in the East, is that of the breath of life.

Since respiration is indeed a central aspect of the metabolism of all but the simplest forms of life, the breath of life seems to be a perfect metaphor for the network of metabolic processes that is the defining characteristic of all living systems. Spirit—the breath of life—is what we have in common with all living beings. It nourishes us and keeps us alive.

Spirituality, or the spiritual life, is usually understood as a way of being that flows from a certain profound experience of reality, which is known as "mystical," "religious," or "spiritual" experience. There are numerous descriptions of this experience in the literature of the world's religions, which tend to agree that it is a direct, nonintellectual experience of reality with some fundamental characteristics that are independent of cultural and historical contexts. One of the most beautiful contemporary descriptions can be found in a short essay titled "Spirituality as Common Sense" by the Benedictine monk, psychologist, and author David Steindl-Rast.[80]

In accordance with the original meaning of spirit as the breath of life, Brother David characterizes spiritual experience as moments of heightened aliveness. Our spiritual moments are those moments when we feel most intensely alive. The aliveness felt during such a "peak experience," as psychologist Abraham Maslow called it, involves not only the body but also the mind. Buddhists refer to this heightened mental alertness as "mindfulness," and they emphasize, interestingly, that mindfulness is deeply rooted in the body. Spirituality, then, is always embodied. We experience our spirit, in the words of Brother David, as "the fullness of mind and body."

It is evident that this notion of spirituality is consistent with the notion of the embodied mind that is now being developed in cognitive science. Spiritual experience is an experience of aliveness of mind and body as a unity. Moreover, this experience of unity transcends not only the separation of mind and body, but also the separation of self and world. The central awareness in these spiritual moments is a profound sense of oneness with all, a sense of belonging to the universe as a whole.[81]

This sense of oneness with the natural world is fully borne out by the new scientific conception of life. As we understand how the roots of life reach deep into basic physics and chemistry, how the unfolding of complexity began long before the formation of the first living cells, and how life has evolved for billions of years by using again and again the same basic patterns and processes, we realize how tightly we are connected with the entire fabric of life.

When we look at the world around us, we find that we are not thrown into chaos and randomness but are part of a great order, a grand symphony of life. Every molecule in our body was once a part of previous bodies—living or nonliving—and will be a part of future bodies. In this sense, our body will not die but will live on, again and again, because life lives on. We share not only life's molecules but also its basic principles of organization with the rest of the living world. And since our mind, too, is embodied, our concepts and metaphors are embedded in the web of life together with our bodies and brains. We belong to the universe, we are at home in it, and this experience of belonging can make our lives profoundly meaningful.

| three |

SOCIAL REALITY

In *The Web of Life* I proposed a synthesis of recent theories of living systems, including insights from nonlinear dynamics, or "complexity theory," as it is popularly known.[1] With the previous two chapters I have laid the groundwork for reviewing this synthesis and extending it to the social domain. My aim, as mentioned in the preface, is to develop a unified, systemic framework for the understanding of biological and social phenomena.

Three Perspectives on Life

The synthesis is based on the distinction between two perspectives on the nature of living systems, which I have called the "pattern perspective" and the "structure perspective," and on their integration by means of a third perspective, the "process perspective." More specifically, I have defined the *pattern of organization* of a living system as the configuration of relationships among the system's components that determines the system's essential characteristics, the *structure* of the system as the material embodiment of its pattern of organization, and the *life process* as the continual process of this embodiment.

I chose the terms "pattern of organization" and "structure" to continue the language used in the theories that form the components of my synthesis.[2] However, in view of the fact that the definition of "structure" in the social sciences is quite different from that in the natural sciences, I shall now modify my terminology and use the more general concepts of *form* and *matter* to accommodate different usages of the term "structure." In this more general terminology, the three perspectives on the nature of living systems correspond to the study of form (or pattern of organization), the study of matter (or material structure), and the study of process.

When we study living systems from the perspective of form, we find that their pattern of organization is that of a self-generating network. From the perspective of matter, the material structure of a living system is a dissipative structure, i.e. an open system operating far from equilibrium. From the process perspective, finally, living systems are cognitive systems in which the process of cognition is closely linked to the pattern of autopoiesis. In a nutshell, this is my synthesis of the new scientific understanding of life.

In the diagram below, I have represented the three perspectives as points in a triangle to emphasize that they are fundamentally interconnected. The form of a pattern of organization can only be recognized if it is embodied in matter, and in living systems this embodiment is an ongoing process. A full understanding of any biological phenomenon must incorporate all three perspectives.

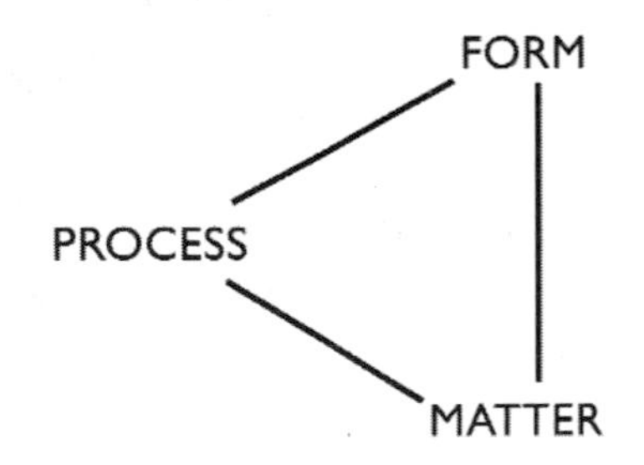

Take, for example, the metabolism of a cell. It consists of a network (*form*) of chemical reactions (*process*), which involve the production of the cell's components (*matter*), and which respond cognitively, i.e. through self-directed structural changes (*process*), to disturbances from the environment. Similarly, the phenomenon of emergence is a *process* characteristic of dissipative structures (*matter*), which involves multiple feedback loops (*form*).

To give equal importance to each of these three perspectives is difficult for most scientists because of the persistent influence of our Cartesian heritage. The natural sciences are supposed to deal with material phenomena, but only one of the three perspectives is concerned with the study of matter. The other two deal with relationships, qualities, patterns, and processes, all of which are nonmaterial. Of course, no scientist would deny the existence of patterns and processes, but most of them think of a pattern as an emergent property of matter, an idea abstracted from matter, rather than a generative force.

To focus on material structures and the forces between them, and to view the patterns of organization resulting from these forces as secondary emergent phenomena has been very effective in physics and chemistry, but when we come to living systems this approach is no longer adequate. The essential characteristic that distinguishes living from nonliving systems—the cellular metabolism—is not a property of matter, nor a special "vital force." It is a specific pattern of relationships among chemical processes.[3] Although it involves relationships between processes that produce material components, the network pattern itself is nonmaterial.

The structural changes in this network pattern are understood as cognitive processes that eventually give rise to conscious experience and conceptual thought. All these cognitive phenomena are nonmaterial, but they are embodied—they arise from and are shaped by the body. Thus, life is never divorced from matter, even though its essential characteristics—organization, complexity, processes, and so on—are nonmaterial.

Meaning—The Fourth Perspective

When we try to extend the new understanding of life to the social domain, we immediately come up against a bewildering multitude of phenomena—rules of behavior, values, intentions, goals, strategies, designs, power relations—that play no role in most of the nonhuman world but are essential to human social life. However, these diverse characteristics of social reality all share a basic common feature, which provides a natural link to the systems view of life developed in the preceding pages.

Self-awareness, as we have seen, emerged during the evolution of our hominid ancestors together with language, conceptual thought, and the social world of organized relationships and culture. Consequently, the understanding of reflective consciousness is inextricably linked to that of language and its social context. This argument can also be turned around: the understanding of social reality is inextricably linked to that of reflective consciousness.

More specifically, our ability to hold mental images of material objects and events seems to be a fundamental condition for the emergence of the key characteristics of social life. Being able to hold mental images enables us to choose among several alternatives, which is necessary to formulate values and social rules of behavior. Conflicts of interest, based on different values, are at the origin of relationships of power, as we shall see below. Our intentions, awareness of purposes and designs and strategies to reach identified goals all require the projection of mental images into the future.

Our inner world of concepts and ideas, images and symbols is a critical dimension of social reality, constituting what John Searle has called "the mental character of social phenomena."[4] Social scientists have often referred to it as the "hermeneutic"* dimension to express the view that human language, being of a symbolic nature, centrally involves the communication of meaning, and that human action flows from the meaning that we attribute to our surroundings.

*From the Greek *hermeneuin* ("to interpret").

Accordingly, I postulate that the systemic understanding of life can be extended to the social domain by adding the perspective of *meaning* to the other three perspectives of life. In doing so, I am using "meaning" as a shorthand notation for the inner world of reflective consciousness, which contains a multitude of interrelated characteristics. A full understanding of social phenomena, then, must involve the integration of four perspectives—form, matter, process, and meaning.

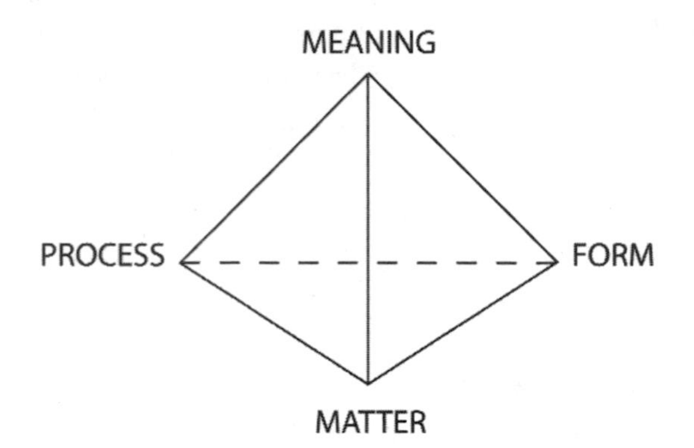

In the diagram above, I have again indicated the interconnectedness of these perspectives by representing them as the corners of a geometric figure. The first three perspectives form a triangle, as before. The perspective of meaning is represented as lying outside the plane of this triangle to indicate that it opens up a new "inner" dimension, so that the entire conceptual structure forms a tetrahedron.

Integrating the four perspectives means recognizing that each contributes significantly to the understanding of a social phenomenon. For example, we shall see that culture is created and sustained by a network (*form*) of communications (*process*), in which *meaning* is generated. The culture's material embodiments (*matter*) include artifacts and written texts, through which meaning is passed on from generation to generation.

It is interesting to note that this conceptual framework of four interdependent perspectives on life shows some similarities with the four principles, or "causes," postulated by Aristotle as the interdependent sources of all phenomena.[5] Aristotle distinguished between internal and external causes. The two internal causes are matter and form. The external causes are the efficient cause, which generates the phenomenon through its action, and the final cause, which determines the action of the efficient cause by giving it a goal or purpose.

Aristotle's detailed description of the four causes and their interrelations is quite different from the conceptual scheme I am proposing.[6] In particular, the final cause, which corresponds to the perspective I have associated with meaning, operates throughout the material world, according to Aristotle, whereas contemporary science asserts that it plays no role in nonhuman systems. Nevertheless, I find it fascinating that after more than 2,000 years of philosophy, we still analyze reality within the four perspectives identified by Aristotle.

Social Theory

When we follow the development of the social sciences from the nineteenth century to the present, we can see that the major debates among different schools of thought seem to reflect the tensions between the four perspectives on social life—form, matter, process, and meaning.

Social thought in the late nineteenth and early twentieth centuries was greatly influenced by positivism, a doctrine formulated by the social philosopher Auguste Comte. Its assertions include the insistence that the social sciences should search for general laws of human behavior, an emphasis on quantification and the rejection of explanations in terms of subjective phenomena, such as intentions or purposes.

It is evident that the positivist framework is patterned after classical physics. Indeed, Auguste Comte, who introduced the term "sociology," first called the scientific study of society "social physics." The major schools of thought in early-twentieth-century sociology can be seen as attempts at emancipation from the positivist straitjacket. In fact, most social theorists of that time positioned themselves explicitly in opposition to the positivist epistemology.[7]

One inheritance of positivism during the early decades of sociology was the focus on a narrow notion of "social causation," which linked social theory conceptually to physics, rather than to the life sciences. Emile Durkheim, who, along with Max Weber, is considered one of the principal founders of modern sociology, identified "social facts," such as beliefs or practices, as the causes of social phenomena. Even though these social facts are clearly nonmaterial, Durkheim insisted that they should be treated like material objects. He saw social facts as being caused by other social facts, in analogy to the operations of physical forces.

Durkheim's ideas exerted a major influence on both structuralism and functionalism, the two dominant schools of early-twentieth-century sociology. Both of these schools of thought assumed that the task of social scientists is to unravel a hidden causative reality beneath the surface level of observed phenomena. Such attempts to identify some hidden phenomena—vital forces or other "extra ingredients"—have occurred repeatedly in the life sciences when scientists struggled to understand the emergence of novelty that is characteristic of all life and cannot be explained in terms of linear relations of cause and effect.

For structuralists, the hidden realm consists of underlying "social structures." Although early structuralists treated those social structures like material objects, they also understood them as integrated wholes and used the term "structure" not unlike the ways in which early systems thinkers used "pattern of organization."

By contrast, the functionalists postulated that there is an underlying social rationality that causes individuals to act according to the "social functions" of their actions—that is, to act in such a way that their actions fulfill society's needs. Durkheim insisted that a full explanation of social phenomena must combine both causal and functional analyses, and he also emphasized that one should distinguish between functions and intentions. It seems that, somehow, he attempted to take into account intentions and purposes (the perspective of *meaning*) without abandoning the conceptual framework of classical physics with its material structures, forces, and linear cause-and-effect relationships.

Several of the early structuralists also recognized the connections between social reality, consciousness, and language. The linguist

Ferdinand de Saussure was one of the founders of structuralism, and the anthropologist Claude Lévi-Strauss, whose name is closely associated with the structuralist tradition, was one of the first to analyze social life by systematically employing analogies with linguistic systems. The focus on language intensified around the 1960s with the advent of the so-called interpretative sociologies, which emphasize that individuals interpret their surrounding reality and act accordingly.

During the 1940s and 1950s, Talcott Parsons, one of the leading social theorists of that time, developed a "general theory of actions" that was heavily influenced by general systems theory. Parsons attempted to integrate structuralism and functionalism into a single theoretical framework, emphasizing that people's actions are both goal-oriented and constrained. Like Parsons, many sociologists of the time introduced the relevance of intentions and purposes by focusing on "human agency," or purposeful action.

The systemic orientation of Talcott Parsons has been advanced further by Niklas Luhmann, one of the most innovative contemporary sociologists, who was inspired by the ideas of Maturana and Varela to develop a theory of "social autopoiesis" to which I shall return in more detail.[8]

Giddens and Habermas—Two Integrative Theories

During the second half of the twentieth century, social theory was shaped significantly by several attempts to transcend the opposing schools of the earlier decades and to integrate the notions of social structure and human agency with an explicit analysis of meaning. The structuration theory of Anthony Giddens and the critical theory of Jürgen Habermas have been perhaps the most influential of those integrative theoretical frameworks.

Anthony Giddens has been a leading contributor to social theory since the early 1970s.[9] His structuration theory is designed to explore the interaction between social structures and human agency in such a way that it integrates insights from structuralism and functionalism on the one hand, and from interpretative sociologies on the other. To do so, Giddens employs two different but complementary methods of investigation. Institutional analysis is his method for studying social structures and institutions, while strategic analysis is used to study how people draw upon social structures in their pursuit of strategic goals.

Giddens emphasizes that people's strategic conduct is based largely on how they interpret their environment. In fact, he points out that social scientists have to deal with a "double hermeneutic." They interpret their subject matter, which itself is engaged in interpretations. Consequently, Giddens believes that subjective phenomenological insights must be taken seriously if we are to understand human conduct.

As would be expected from an integrative theory that attempts to transcend traditional opposites, Giddens's concept of social structure is rather complex. As in most contemporary social theory, it is defined as a set of rules enacted in social practices, and Giddens also includes resources in his definition of social structure. The rules are of two kinds: interpretative schemes, or semantic rules; and norms, or moral rules. There are also two kinds of resources. Material resources include the ownership or control of objects (the traditional focus of Marxist sociologies), while authoritative resources result from the organization of power.

Giddens also uses the terms "structural properties" for the institutionalized features of society (e.g., the division of labor) and "structural principles" for the most deeply embedded of those features. The study of structural principles, the most abstract form of social analysis, allows one to distinguish between different types of societies.

The interaction between social structures and human agency is cyclical, according to Giddens. Social structures are both the precondition and the unintended outcome of people's agency. People draw upon them in order to engage in their daily social practices, and in so doing they cannot help but reproduce the very same structures.

For example, when we speak we necessarily draw upon the rules of our language, and as we use language we continually reproduce and transform the very same semantic structures. Thus social structures both enable us to interact and are also reproduced by our interactions. Giddens calls this the "duality of structure," and he acknowledges the similarity to the circular nature of autopoietic networks in biology.[10]

The conceptual links with the theory of autopoiesis are even more evident when we turn to Giddens's view of human agency. He insists that agency does not consist of discrete acts but is a continuous flow of conduct. Similarly, a living metabolic network embodies an ongoing process of life. And as the components of the living network continually transform or replace other components, so the actions in the flow of human conduct have a "transformative capacity" in Giddens's theory.

During the 1970s, while Anthony Giddens developed his structuration theory at Cambridge University, Jürgen Habermas formulated a theory of equal scope and depth, which he called the "theory of communicative action," at the University of Frankfurt.[11] By integrating numerous philosophical strands, Habermas has become a leading intellectual force and a major influence on philosophy and social theory. He is the most prominent contemporary exponent of critical theory, the social theory with Marxist roots that was developed by the Frankfurt School in the 1930s.[12] True to their Marxist origins, critical theorists do not simply want to explain the world. Their ultimate task, according to Habermas, is to uncover the structural conditions of people's actions and to help them transcend these conditions. Critical theory deals with power and is aimed at emancipation.

Like Giddens, Habermas asserts that two different but complementary perspectives are needed to fully understand social phenomena. One perspective is that of the social system, which corresponds to the focus on institutions in Giddens's theory; the other is the perspective of the "life-world" (*Lebenswelt*), corresponding to Giddens's focus on human conduct.

For Habermas, the social system has to do with the ways social structures constrain people's actions, which includes issues of power and specifically the class relationships involved in production. The life-world, on the other hand, raises issues of meaning and communication. Accordingly, Habermas sees critical theory as the integration of two different types of knowledge. Empirical-analytical knowledge is associated with the external world and is concerned with causal explanations. Hermeneutics, the understanding of meaning, is associated with the inner world, and is concerned with language and communication.

Like Giddens, Habermas recognizes that hermeneutic insights are relevant to the workings of the social world because people attribute meaning to their surroundings and act accordingly. However, he points out that people's interpretations always rely on a number of implicit assumptions that are embedded in history and tradition, and he argues that this means that all assumptions are not equally valid. According to Habermas, social scientists should evaluate different traditions critically, identify ideological distortions, and uncover their connections with power relations. Emancipation takes place whenever people are able to overcome past restrictions that resulted from distorted communication.

In accordance with his distinctions between different worlds and types of knowledge, Habermas also distinguishes between different types of action, and here the integrative nature of his critical theory is perhaps most evident. In terms of the four perspectives on life introduced above, we can say that action clearly belongs to the process perspective. By identifying three types of action, Habermas connects *process* with each of the other three perspectives. Instrumental action takes place in the external world (*matter*); strategic action deals with human relationships (*form*); and communicative action is oriented toward reaching understanding (*meaning*). Each type of action is associated with a different sense of "rightness" for Habermas. Right action refers to factual truth in the material world, to moral rightness in the social world, and to sincerity in the inner world.

Extending the Systems Approach

The theories of Giddens and Habermas are outstanding attempts to integrate studies of the external world of cause and effect, the social world of human relationships, and the inner world of values and meaning. Both social theorists integrate insights from the natural sciences, the social sciences and from cognitive philosophies, while rejecting the limitations of positivism.

I believe that this integration can be advanced significantly by extending the new systemic understanding of life to the social domain within the conceptual framework of the four perspectives introduced above—form, matter, process, and meaning. We need to integrate all four perspectives to reach a systemic understanding of social reality.

Such a systemic understanding is based on the assumption that there is a fundamental unity to life, that different living systems exhibit similar patterns of organization. This assumption is supported by the observation that evolution has proceeded for billions of years by using the same patterns again and again. As life evolves, these patterns tend to become more and more elaborate, but they are always variations on the same basic themes.

The network, in particular, is one of the very basic patterns of organization in all living systems. At all levels of life—from the metabolic networks of cells to the food webs of ecosystems—the components and processes of living systems are interlinked in network fashion. Extending the systemic understanding of life to the social domain, therefore, means applying our knowledge of life's basic patterns and principles of organization, and specifically our understanding of living networks, to social reality.

However, while insights into the organization of biological networks may help us understand social networks, we should not expect to transfer our understanding of the network's material structure from the biological to the social domain. Let us take the metabolic network of cells as an example to illustrate this point. A cellular network is a nonlinear pattern of organization, and we need complexity theory (nonlinear dynamics) to understand its intricacies. The cell, moreover, is a chemical system, and we need molecular biology and biochemistry to understand the nature of the structures and processes that form the network's nodes and links. If we do not know what an enzyme is and how it catalyzes the synthesis of a protein, we cannot expect to understand the cell's metabolic network.

A social network, too, is a nonlinear pattern of organization, and concepts developed in complexity theory, such as feedback or emergence, are likely to be relevant in a social context as well, but the nodes and links of the network are not merely biochemical. Social networks are first and foremost networks of communication involving symbolic language, cultural constraints, relationships of power, and so on. To understand the structures of such networks we need to use insights from social theory, philosophy, cognitive science, anthropology, and other disciplines. A unified systemic framework for the understanding of biological and social phenomena will emerge only when the concepts of nonlinear dynamics are combined with insights from these fields of study.

Networks of Communications

To apply our knowledge of living networks to social phenomena, we need to find out whether the concept of autopoiesis is valid in the social domain. There has been considerable discussion of this point in recent years, but the situation is still far from clear.[13] The key question is: What are the elements of an autopoietic social network? Maturana and Varela originally proposed that the concept of autopoiesis should be restricted to the description of cellular networks, and that the broader concept of "organizational closure," which does not specify production processes, should be applied to all other living systems.

Another school of thought, pioneered by sociologist Niklas Luhmann, holds that the notion of autopoiesis *can* be extended to the social domain and formulated strictly within the conceptual framework of social theory. Luhmann has developed a theory of "social autopoiesis" in considerable detail.[14] However, he takes the curious position that social systems, while being autopoietic, are not living systems.

Since social systems not only involve living human beings, but also language, consciousness, and culture, they are evidently cognitive systems—it seems rather strange to consider them as not being alive. I prefer to retain autopoiesis as a defining characteristic of life, but in my discussion of human organizations I will also suggest that social systems can be alive to varying degrees.[15]

Luhmann's central point is to identify communications as the elements of social networks: "Social systems use communication as their particular mode of autopoietic reproduction. Their elements are communications that are recursively produced and reproduced by a network of communications and that cannot exist outside of such a network."[16] These networks of communications are self-generating. Each communication creates thoughts and meaning, which give rise to further communications, and thus the entire network generates itself—it is autopoietic. As communications recur in multiple feedback loops, they produce a shared system of beliefs, explanations, and values—a common context of meaning—that is continually sustained by further communications. Through this shared context of meaning individuals acquire identities as members of the social network, and in this way the network generates its own boundary. It is not a physical boundary but a boundary of expectations, of confidentiality and loyalty, which is continually maintained and renegotiated by the network itself.

To explore the implications of viewing social systems as networks of communications, it is helpful to remember the dual nature of human communication. Like all communication among living organisms, it involves a continual coordination of behavior, and because it involves conceptual thinking and symbolic language it also generates mental images, thoughts, and meaning. Accordingly, we can expect networks of communications to have a dual effect. They will generate, on the one hand, ideas and contexts of meaning, and on the other hand, rules of behavior or, in the language of social theorists, social structures.

Meaning, Purpose, and Human Freedom

Having identified the organization of social systems as self-generating networks, we now need to turn our attention to the structures that are produced by these networks and to the nature of the relationships that are engendered by them. A comparison with biological networks will again be useful. The metabolic network of a cell, for example, generates material structures. Some of them become structural components of the network, forming parts of the cell membrane or of other cellular structures. Others are exchanged between the network's nodes as carriers of energy or information, or as catalysts of metabolic processes.

Social networks, too, generate material structures—buildings, roads, technologies, etc.—that become structural components of the network; and they also produce material goods and artifacts that are exchanged between the network's nodes. However, the production of material structures in social networks is quite different from that in biological and ecological networks. The structures are created for a purpose, according to some design, and they embody some meaning. To understand the activities of social systems, it is crucial to study them from that perspective.

The perspective of meaning includes a multitude of interrelated characteristics that are essential to understanding social reality. Meaning itself is a systemic phenomenon: it always has to do with context. *Webster's Dictionary* defines meaning as "an idea conveyed to the mind that requires or allows of interpretation," and interpretation as "conceiving in the light of individual belief, judgment, or circumstance." In other words, we interpret something by putting it into a particular context of concepts, values, beliefs, or circumstances. To understand the meaning of anything we need to relate it to other things in its environment, in its past, or in its future. Nothing is meaningful in itself.

For example, to understand the meaning of a literary text, one needs to establish the multiple contexts of its words and phrases. This can be a purely intellectual endeavor, but it may also reach a deeper level. If the context of an idea or expression includes relationships involving our own selves, it becomes meaningful to us in a personal way. This deeper sense of meaning includes an emotional dimension and may even bypass reason altogether. Something may be profoundly meaningful to us through context provided by direct experience.

Meaning is essential to human beings. We continually need to make sense of our outer and inner worlds, find meaning in our environment and in our relationships with other humans, and act according to that meaning. This includes in particular our need to act with a purpose or goal in mind. Because of our ability to project mental images into the future we act with the conviction, valid or invalid, that our actions are voluntary, intentional, and purposeful.

As human beings we are capable of two kinds of actions. Like all living organisms we engage in involuntary, unconscious activities, such as digesting our food or circulating our blood, which are part of the process of life and therefore cognitive in the sense of the Santiago Theory. In addition, we engage in voluntary, intentional activities, and it is in acting with intention and purpose that we experience human freedom.[17]

As I mentioned above, the new understanding of life sheds new light on the age-old philosophical debate about freedom and determinism.[18] The key point is that the behavior of a living organism is constrained but not determined by outside forces. Living organisms are self-organizing, which means that their behavior is not imposed by the environment but is established by the system itself. More specifically, the organism's behavior is determined by its own structure, a structure formed by a succession of autonomous structural changes.

The autonomy of living systems must not be confused with independence. Living organisms are not isolated from their environment. They interact with it continually, but the environment does not determine their organization. At the human level, we experience this self-determination as the freedom to act according to our own choices and decisions. To experience these as our own means that they are determined by our nature, including our past experiences and genetic heritage. To the extent that we are not constrained by human relationships of power, our behavior is self-determined and therefore free.

The Dynamics of Culture

Our ability to hold mental images and project them into the future not only allows us to identify goals and purposes and develop strategies and designs, but also enables us to choose among several alternatives and hence to formulate values and social rules of behavior. All of these social phenomena are generated by networks of communications as a consequence of the dual role of human communication. On the one hand, the network continually generates mental images, thoughts, and meaning; on the other hand, it continually coordinates the behavior of its members. From the complex dynamics and interdependence of these processes emerges the integrated system of values, beliefs, and rules of conduct that we associate with the phenomenon of culture.

The term "culture" has a long and intricate history and is now used in different intellectual disciplines with diverse and sometimes confusing meanings. In his classic text, *Culture*, historian Raymond Williams traces the meaning of the word back to its early use as a noun denoting a process: the culture (i.e. cultivation) of crops, or the culture (i.e. rearing and breeding) of animals. In the sixteenth century this meaning was extended metaphorically to the active cultivation of the human mind; and in the late eighteenth century, when the word was borrowed from the French by German writers (who first spelled it *Cultur* and subsequently *Kultur*), it acquired the meaning of a distinctive way of life of a people.[19] In the nineteenth century the plural "cultures" became especially important in the development of comparative anthropology, where it has continued to designate distinctive ways of life.

In the meantime, the older use of "culture" as the active cultivation of the mind continued. Indeed, it expanded and diversified, covering a range of meanings from a developed state of mind ("a cultured person") to the process of this development ("cultural activities") to the means of these processes (administered, for example, by a "Ministry of Culture"). In our time, the different meanings of "culture" that are associated with the active cultivation of the mind coexist—often uneasily, as Williams notes—with the anthropological use as a distinctive way of life of a people or social group (as in "aboriginal culture" or "corporate culture"). In addition, the original biological meaning of "culture" as cultivation continues to be used, as for example in "agriculture," "monoculture," or "germ culture."

For our systemic analysis of social reality we need to focus on the anthropological meaning of culture, which the *Columbia Encyclopedia* defines as "the integrated system of socially acquired values, beliefs, and rules of conduct that delimit the range of accepted behaviors in any given society." When we explore the details of this definition, we discover that culture arises from a complex, highly nonlinear dynamic. It is created by a social network involving multiple feedback loops through which values, beliefs, and rules of conduct are continually communicated, modified, and sustained. It emerges from a network of communications among individuals; and as it emerges, it produces constraints on their actions. In other words, the social structures, or rules of behavior, that constrain the actions of individuals are produced and continually reinforced by their own network of communications.

The social network also produces a shared body of knowledge—including information, ideas, and skills—that shapes the culture's distinctive way of life in addition to its values and beliefs. Moreover, the culture's values and beliefs affect its body of knowledge. They are part of the lens through which we see the world. They help us to interpret our experiences and to decide what kind of knowledge is meaningful. This meaningful knowledge, continually modified by the network of communications, is passed on from generation to generation together with the culture's values, beliefs, and rules of conduct.

The system of shared values and beliefs creates an identity among the members of the social network, based on a sense of belonging. People in different cultures have different identities because they share different sets of values and beliefs. At the same time, an individual may belong to several different cultures. People's behavior is informed and restricted by their cultural identities, which in turn reinforces their sense of belonging. Culture is embedded in people's way of life, and it tends to be so pervasive that it escapes our everyday awareness.

Cultural identity also reinforces the closure of the network by creating a boundary of meaning and expectations that limits the access of people and information to the network. Thus the social network is engaged in communication within a cultural boundary which its members continually re-create and renegotiate. This situation is not unlike that of the metabolic network of a cell, which continually produces and re-creates a boundary—the cell membrane—that confines it and gives it its identity. However, there are some crucial differences between cellular and social boundaries. Social boundaries, as I have emphasized, are not necessarily physical boundaries but boundaries of meaning and expectations. They do not literally surround the network, but exist in a mental realm that does not have the topological properties of physical space.

The Origin of Power

One of the most striking characteristics of social reality is the phenomenon of power. In the words of economist John Kenneth Galbraith, "The exercise of power, the submission of some to the will of others, is inevitable in modern society; nothing whatever is accomplished without it . . . Power can be socially malign; it is also socially essential."[20] The essential role of power in social organization is linked to inevitable conflicts of interest. Because of our ability to affirm preferences and make choices accordingly, conflicts of interest will appear in any human community, and power is the means by which these conflicts are resolved.

This does not necessarily imply the threat or use of violence. In his lucid essay, Galbraith distinguishes three kinds of power, depending on the means that are employed. Coercive power wins submission by inflicting or threatening sanctions; compensatory power by offering incentives or rewards; and conditioned power by changing beliefs through persuasion or education.[21] To find the right mixture of these three kinds of power in order to resolve conflicts and balance competing interests is the art of politics.

Relationships of power are culturally defined by agreements on positions of authority that are part of the culture's rules of conduct. In human evolution, such agreements may have emerged very early on with the development of the first communities. A community would be able to act much more effectively if somebody had the authority to make or facilitate decisions when there were conflicts of interest. Such social arrangements would have given the community a significant evolutionary advantage.

Indeed, the original meaning of "authority" is not "power to command," but "a firm basis for knowing and acting."[22] When we need a firm basis for knowing, we might consult an authoritative text; when we have a serious illness, we look for a doctor who is an authority in the relevant field of medicine.

From the earliest times, human communities have chosen men and women as their leaders when they recognized their wisdom and experience as a firm basis for collective action. These leaders were then invested with power, which meant originally that they were given ritual vestments as symbols of their leadership, and their authority became associated with the power to command. The origin of power, then, lies in culturally defined positions of authority on which the community relies for the resolution of conflicts and for decisions about how to act wisely and effectively. In other words, true authority consists in empowering others to act.

However, it often happens that the vestment that gives the power to command—the piece of cloth, crown, or other symbol—is passed on to someone without true authority. This invested authority, rather than the wisdom of a genuine leader, is now the only source of power, and in this situation its nature can easily change from empowering others to the advancement of an individual's own interests. This is when power becomes linked to exploitation.

The association of power with the advancement of one's own interests is the basis of most contemporary analyses of power. In the words of Galbraith, "Individuals and groups seek power to advance their own interests and to extend to others their personal, religious, or social values."[23] A further stage of exploitation is reached when power is pursued for its own sake. It is well known that for most people the exercise of power brings high emotional and material rewards, conveyed by elaborate symbols and rituals of obeisance—from standing ovations, fanfares, and military salutes to office suites, limousines, corporate jets, and motorcades.

As a community grows and increases in complexity, its positions of power will also increase. In complex societies, resolutions of conflicts and decisions about how to act will be effective only if authority and power are organized within administrative structures. In the long history of human civilization, numerous forms of social organization have been generated by this need to organize the distribution of power.

Thus, power plays a central role in the emergence of social structures. In social theory, all rules of conduct are included in the concept of social structures, whether they are informal, resulting from continual coordinations of behavior, or formalized, documented, and enforced by laws. All such formal structures, or social institutions, are ultimately rules of behavior that facilitate decision-making and embody relationships of power. This crucial link between power and social structure has been discussed extensively in the classic texts on power. Sociologist and economist Max Weber states: "Domination has played the decisive role . . . in the economically most important social structures of the past and present";[24] and according to political theorist Hannah Arendt: "All political institutions are manifestations and materializations of power."[25]

Structure in Biological and Social Systems

As we explored the dynamics of social networks, of culture, and of the origin of power in the preceding pages, we saw repeatedly that the generation of structures, both material and social, is a key characteristic of those dynamics. At this point, it is useful to review the role of structure in living systems in a systematic way.

The central focus of a systemic analysis is the notion of organization, or "pattern of organization." Living systems are self-generating networks, which means that their pattern of organization is a network pattern in which each component contributes to the production of other components. This idea can be extended to the social domain by identifying the relevant living networks as networks of communications.

In the social realm, the concept of organization takes on an additional meaning. Social organizations, such as businesses or political institutions, are systems whose patterns of organization are designed specifically to distribute power. These formally designed patterns are known as organizational structures and are visually represented by the standard organizational charts. They are ultimately rules of behavior that facilitate decision-making and embody relationships of power.[26]

In biological systems, all structures are material structures. The processes in a biological network are production processes of the network's material components, and the resulting structures are the material embodiments of the system's pattern of organization. All biological structures change continually; so the process of material embodiment is continual.

Social systems produce nonmaterial as well as material structures. The processes that sustain a social network are processes of communication, which generate shared meaning and rules of behavior (the network's culture), as well as a shared body of knowledge. The rules of behavior, whether formal or informal, are called social structures. Sociologist Manuel Castells states that: "Social structures are the foundational concept of social theory. Everything else works through the social structures."[27]

The ideas, values, beliefs, and other forms of knowledge generated by social systems constitute structures of meaning, which I shall call "semantic structures." These semantic structures, and thus the network's patterns of organization, are embodied physically to some extent in the brains of the individuals belonging to the network. They may also be embodied in other biological structures through the effects of people's minds on their bodies, as, for example, in stress-related illnesses. Recent discoveries in cognitive science imply that, since the mind is always embodied, there is continual interplay between semantic, neural, and other biological structures.[28]

In modern societies, the culture's semantic structures are documented—that is, materially embodied—in written and digital texts. They are also embodied in artifacts, works of art, and other material structures, as they are in traditional nonliterate cultures. Indeed, the activities of individuals in social networks specifically include the organized production of material goods. All these material structures—texts, works of art, technologies, and material goods—are created for a purpose and according to some design. They are embodiments of the shared meaning generated by the society's networks of communications.

Technology and Culture

In biology, the behavior of a living organism is shaped by its structure. As the structure changes during the organism's development and during the evolution of its species, so does its behavior.[29] A similar dynamic can be observed in social systems. The biological structure of an organism corresponds to the material infrastructure of a society, which embodies the society's culture. As the culture evolves, so does its infrastructure—they coevolve through continual mutual influences.

The influences of the material infrastructure on people's behavior and culture are especially significant in the case of technology, hence the analysis of technology has become an important subject in social theory, both within and beyond the Marxist tradition.[30]

The meaning of "technology," like that of "science," has changed considerably over the centuries. The original Greek *technologia*, derived from *techne* ("art"), meant a discourse on the arts. When the term was first used in English in the seventeenth century, it meant a systematic discussion of the "applied arts," or crafts, and gradually it came to denote the crafts themselves. In the early twentieth century, the meaning was extended to include not only tools and machines but also nonmaterial methods and techniques, meaning a systematic application of any such techniques. Thus, we speak of "the technology of management," or of "simulation technologies." Today, most definitions of technology emphasize its connection with science. Sociologist Manuel Castells defines technology as "the set of tools, rules, and procedures through which scientific knowledge is applied to a given task in a reproducible manner."[31]

Technology, however, is much older than science. Its origins in toolmaking go back to the very dawn of the human species when language, reflective consciousness, and the ability to make tools evolved together.[32] Accordingly, the first human species was given the name *homo habilis* ("skillful human") to denote its ability to make sophisticated tools.[33] Technology is a defining characteristic of human nature: its history encompasses the entire history of human evolution.

Being a fundamental aspect of human nature, technology has crucially shaped successive epochs of civilization.[34] We characterize the great periods of human civilization in terms of their technologies—from the Stone Age, Bronze Age, and Iron Age to the Industrial Age and the Information Age. Throughout the millennia, but especially since the Industrial Revolution, critical voices have pointed out that the influences of technology on human life and culture are not always beneficial. In the early nineteenth century, William Blake decried the "dark Satanic mills" of Great Britain's growing industrialism, and several decades later Karl Marx vividly and movingly described the horrendous exploitation of workers in the British lace and pottery industries.[35]

More recently, critics have emphasized the increasing tensions between cultural values and high technology.[36] Technology advocates often discount those critical voices by claiming that technology is neutral: that it can have beneficial or harmful effects depending on how it is used. However, these defenders of technology do not realize that a specific technology will always shape human nature in specific ways, because the use of technology is such a fundamental aspect of being human. As historians Melvin Kranzberg and Carroll Pursell explain:

> To say that technology is not strictly neutral, that it has inherent tendencies or imposes its own values, is merely to recognize the fact that, as a part of our culture, it has an influence on the way in which we behave and grow. Just as [humans] have always had some form of

> technology, so has that technology influenced the nature and direction of their development. The process cannot be stopped nor the relationship ended; it can only be understood and, hopefully, directed toward goals worthy of [humankind].[37]

This brief discussion of the interplay between technology and culture, to which I shall return several times in the subsequent pages, concludes my outline of a unified, systemic framework for the understanding of biological and social life. In the remainder of this book, I shall apply this new conceptual framework to some of the most critical social and political issues of our time—the management of human organizations, the challenges and dangers of economic globalization, the problems of biotechnology and the design of sustainable communities.

| part two |

THE CHALLENGES OF THE TWENTY-FIRST CENTURY

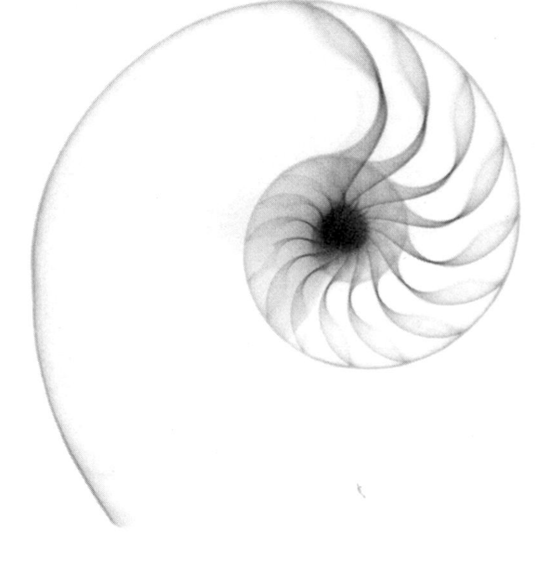

| four |

LIFE AND LEADERSHIP IN ORGANIZATIONS

In recent years, the nature of human organizations has been discussed extensively in business and management circles in response to a widespread feeling that today's businesses need to undergo fundamental transformations. Organizational change has become a dominant theme in management literature, and numerous business consultants offer seminars on "change management."

Over the past ten years, I have been invited to speak at quite a few business conferences, and at first I was very puzzled when I encountered the strongly felt need for organizational change. Corporations seemed to be more powerful than ever; business was clearly dominating politics; and the profits and shareholder values of most companies were rising to unprecedented heights. Things seemed to be going very well indeed for business, so why was there so much talk about fundamental change?

As I listened to the conversations among business executives at these seminars, I soon began to see a different picture. Top executives are under enormous stress today. They work longer hours than ever before, and many of them complain that they have no time for personal relationships and experience little satisfaction in their lives in spite of increasing material prosperity. Their companies may look powerful from outside, but they themselves feel pushed around by global market forces and insecure in the face of turbulence they can neither predict nor fully comprehend.

The business environment of most companies today changes with incredible speed. Markets are rapidly being deregulated, and never-ending corporate mergers and acquisitions impose radical cultural and structural changes on the organizations involved—changes that go beyond people's learning capabilities and overwhelm both individuals and organizations. As a result, there is a deep and pervasive feeling among managers that, no matter how hard they work, things are out of control.

Complexity and Change

The root cause of this deep malaise among business executives seems to be the enormous complexity that has become one of the foremost characteristics of present-day industrial society. At the beginning of this new century, we are surrounded by massively complex systems that increasingly permeate almost every aspect of our lives. These complexities were difficult to imagine only half a century ago—global trading and broadcast systems, instant worldwide communication via ever more sophisticated electronic networks, giant multinational organizations, automated factories, and so on.

The amazement we feel in contemplating these wonders of industrial and informational technologies is tinged by a sense of uneasiness, if not outright discomfort. Even though these complex systems continue to be hailed for their increasing sophistication, there is a growing recognition that they have brought with them a business and organizational environment that is almost unrecognizable from the point of view of traditional management theory and practice.

As if that were not alarming enough, it is becoming ever more apparent that our complex industrial systems, both organizational and technological, are the main driving force of global environmental destruction, and the main threat to the long-term survival of humanity. To build a sustainable society for our children and future generations, we need to fundamentally redesign many of our technologies and social institutions so as to bridge the wide gap between human design and the ecologically sustainable systems of nature.[1]

Organizations need to undergo fundamental changes, both in order to adapt to the new business environment and to become ecologically sustainable. This double challenge is urgent and real, and the recent extensive discussions of organizational change are fully justified. However, despite these discussions and some anecdotal evidence of successful attempts to transform organizations, the overall track record is very poor. In recent surveys, CEOs reported again and again that their efforts at organizational change did not yield the promised results. Instead of managing new organizations, they ended up managing the unwanted side effects of their efforts.[2]

At first glance, this situation seems paradoxical. When we look around our natural environment, we see continuous change, adaptation, and creativity; and yet, our business organizations seem to be incapable of dealing with change. Over the years, I have come to realize that the roots of this paradox lie in the dual nature of human organizations.[3] On the one hand, they are social institutions designed for specific purposes, such as making money for their shareholders, managing the distribution of political power, transmitting knowledge, or spreading religious faith. At the same time, organizations are communities of people who interact with one another to build relationships, help each other, and make their daily activities meaningful at a personal level.

These two aspects of organizations correspond to two very different types of change. Many CEOs are disappointed about their efforts to achieve change in large part because they see their company as a well-designed tool for achieving specific purposes, and when they attempt to change its design they want predictable, quantifiable change in the entire structure. However, the designed structure always intersects with the organization's living individuals and communities, for whom change cannot be designed.

It is common to hear that people in organizations resist change. In reality, people do not resist change; they resist having change imposed on them. Being alive, individuals and their communities are both stable and subject to change and development, but their natural change processes are very different from the organizational changes designed by "reengineering" experts and mandated from the top.

To resolve the problem of organizational change, we first need to understand the natural change processes that are embedded in all living systems. Once we have that understanding, we can begin to design the processes of organizational change accordingly and to create human organizations that mirror life's adaptability, diversity, and creativity.

According to the systemic understanding of life, living systems continually create, or re-create, themselves by transforming or replacing their components. They undergo continual structural changes while preserving their weblike patterns of organization.[4] Understanding life means understanding its inherent change processes. It seems that organizational change will appear in a new light when we understand clearly to what extent and in what ways human organizations are alive. As organizational theorists Margaret Wheatley and Myron Kellner-Rogers put it, "Life is the best teacher about change."[5]

What I am proposing, following Wheatley and Kellner-Rogers, is a systemic solution to the problem of organizational change, which, like many systemic solutions, solves not only that problem but also several others. Understanding human organizations in terms of living systems, i.e. in terms of complex nonlinear networks, is likely to lead to new insights into the nature of complexity, and thus help us deal with the complexities of today's business environment.

Moreover, it will help us design business organizations that are ecologically sustainable, since the principles of organization of ecosystems, which are the basis of sustainability, are identical to the principles of organization of all living systems. It would seem, then, that understanding human organizations as living systems is one of the critical challenges of our time.

There is an additional reason why the systemic understanding of life is of paramount importance in the management of today's business organizations. Over the last few decades we have seen the emergence of a new economy that is shaped decisively by information and communication technologies, and in which the processing of information and creation of scientific and technical knowledge are the main sources of productivity.[6] According to classical economic theory, the key sources of wealth are natural resources (land in particular), capital, and labor. Productivity results from the effective combination of these three sources through management and technology. In today's economy, both management and technology are critically linked to knowledge creation. Increases in productivity do not come from labor, but from the capacity to equip labor with new capabilities, based on new knowledge. Thus "knowledge management," "intellectual capital," and "organizational learning" have become important new concepts in management theory.[7]

According to the systems view of life, the spontaneous emergence of order and the dynamics of structural coupling, which results in the continual structural changes that are characteristic of all living systems, are the basic phenomena underlying the process of learning.[8] Moreover, we have seen that the creation of knowledge in social networks is a key characteristic of the dynamics of culture.[9] Combining these insights and applying them to organizational learning enables us to clarify the conditions under which learning and knowledge creation take place and derive important guidelines for the management of today's knowledge-oriented organizations.

Metaphors in Management

The basic idea of management, underlying both its theory and practice, is that of steering an organization in a direction consistent with its goals and purposes.[10] For business organizations, these prominently include financial goals, and thus, as management theorist Peter Block points out, the chief concerns of management are the definition of purpose, the use of power, and the distribution of wealth.[11]

In order to steer an organization effectively, managers need to know in some detail how it functions, and since the relevant processes and patterns of organization can be very complex, especially in today's large corporations, managers have traditionally used metaphors to identify broad overall perspectives. Organizational theorist Gareth Morgan has analyzed the key metaphors used to describe organizations in an illuminating book, *Images of Organization.* According to Morgan, "The medium of organization and management is metaphor. Management theory and practice is shaped by a metaphorical process that influences virtually everything we do."[12]

The key metaphors he discusses include organizations as machines (with the focus on control and efficiency), as organisms (development, adaptation), as brains (organizational learning), as cultures (values, beliefs), and as systems of government (conflicts of interest, power). From the point of view of our conceptual framework, we see that the organism and brain metaphors address the biological and cognitive dimensions of life respectively, while the culture and government metaphors represent various aspects of the social dimension. The main contrast is between the metaphor of organizations as machines and that of organizations as living systems.

My intent is to go beyond the metaphorical level and see to what extent human organizations can literally be understood as living systems. Before doing so, however, it will be useful to review the history and main characteristics of the machine metaphor. It is an integral part of the much broader mechanistic paradigm that was formulated by Descartes and Newton in the seventeenth century and has dominated our culture for several hundred years, during which it has shaped modern Western society and has significantly influenced the rest of the world.[13]

The view of the universe as a mechanical system composed of elementary building blocks has shaped our perception of nature, of the human organism, of society, and thus also of the business organization. The first mechanistic theories of management were the classical management theories of the early twentieth century, in which organizations were designed as assemblages of precisely interlocking parts—functional departments such as production, marketing, finance, and personnel—linked together through clearly defined lines of command and communication.[14]

This view of management as engineering, based on precise technical design, was perfected by Frederick Taylor, an engineer whose "principles of scientific management" provided the cornerstone of management theory during the first half of the twentieth century. As Gareth Morgan points out, Taylorism in its original form is still alive in numerous fast-food chains around the world. In these mechanized restaurants that serve hamburgers, pizzas, and other highly standardized products,

> work is often organized in the minutest detail on the basis of designs that analyse the total process of production, find the most efficient procedures, and then allocate these as specialized duties to people trained to perform them in a very precise way. All the thinking is done by the managers and designers, leaving all the doing to the employees.[15]

The principles of classical management theory have become so deeply ingrained in the ways we think about organizations that for most managers the design of formal structures, linked by clear lines of communication, coordination, and control, has become almost second nature. We shall see that this largely unconscious embrace of the mechanistic approach to management is one of the main obstacles to organizational change today.

To appreciate the profound impact of the machine metaphor on the theory and practice of management, let us now contrast it with the view of organizations as living systems, still at the level of metaphor for the time being. Management theorist Peter Senge, who has been one of the main proponents of systems thinking and of the idea of the "learning organization" in American management circles, has put together an impressive list of implications of these two metaphors for organizations. To heighten the contrast between them, Senge characterizes one as a "machine for making money" and the other as a "living being."[16]

A machine is designed by engineers for a specific purpose and is owned by someone who is free to sell it. This exactly expresses the mechanistic view of organizations. It implies that a company is created and owned by people outside the system. Its structure and goals are designed by management or by outside experts and are imposed on the organization. If we see the organization as a living being, however, the question of ownership becomes problematic. "Most people in the world," Senge notes, "would regard the idea that one person owns another as fundamentally immoral."[17] If organizations were truly living communities, buying and selling them would be the equivalent of slavery, and subjecting the lives of their members to predetermined goals would be seen as dehumanizing.

To run properly, a machine must be controlled by its operators, so that it will function according to their instructions. Accordingly, the whole thrust of classical management theory is to achieve efficient operations through top-down control. Living beings, on the other hand, act autonomously. They can never be controlled like machines. To try and do so is to deprive them of their aliveness.

Seeing a company as a machine also implies that it will eventually run down, unless it is periodically serviced and rebuilt by management. It cannot change by itself; all changes need to be designed by someone else. To see the company as a living being, by contrast, is to realize that it is capable of regenerating itself and that it will naturally change and evolve.

"The machine metaphor is so powerful," Senge concludes, "that it shapes the character of most organizations. They become more like machines than living beings because their members *think* of them that way."[18] The mechanistic approach to management has certainly been very successful in increasing efficiency and productivity, but it has also resulted in widespread animosity toward organizations that are managed in machinelike ways. The reason for that is obvious. Most people resent being treated like cogs in a machine.

When we look at the contrast between the two metaphors—machine versus living being—it is evident why a management style guided by the machine metaphor will have problems with organizational change. The need to have all changes designed by management and imposed upon the organization tends to generate bureaucratic rigidity. There is no room for flexible adaptations, learning, and evolution in the machine metaphor, and it is clear that organizations managed in strictly mechanistic ways cannot survive in today's complex, knowledge-oriented and rapidly changing business environment.

Peter Senge published his juxtaposition of the two metaphors in a foreword to a remarkable book, titled *The Living Company.*[19] It's author, Arie de Geus, a former Shell executive, approached the question of the nature of business organizations from an interesting angle. In the 1980s, De Geus directed a study for the Shell Group to examine the question of corporate longevity. He and his colleagues looked at large corporations that had existed for over a hundred years, had survived major changes in the world around them, and were still flourishing with their corporate identities intact.

The study analyzed twenty-seven such long-lived corporations and found that they had several key characteristics in common.[20] This led De Geus to conclude that resilient, long-lived companies are those that exhibit the behavior and certain characteristics of living entities. Essentially, he identifies two sets of characteristics. One is a strong sense of community and collective identity around a set of common values; a community in which all members know that they will be supported in their endeavors to achieve their own goals. The other set of characteristics is openness to the outside world, tolerance for the entry of new individuals and ideas, and consequently a manifest ability to learn and adapt to new circumstances.

He contrasts the values of such a learning company, whose main purpose is to survive and thrive in the long run, with those of a conventional "economic company," whose priorities are determined by purely economic criteria. He asserts that "the sharp difference between these two definitions of a company—the economic company definition and the learning company definition—lies at the core of the crisis managers face today."[21] To overcome the crisis, he suggests, managers need to "shift their priorities, from managing companies to optimize capital to managing companies to optimize people."[22]

Social Networks

For De Geus, it does not matter very much whether the "living company" is simply a useful metaphor, or whether business organizations are actually living systems, as long as managers think of a company as being alive and change their management style accordingly. He also urges them to choose between the two images of the "living company" and the "economic company," which seems rather artificial. A company is certainly a legal and economic entity, and in some sense it also seems to be alive. The challenge is to integrate these two aspects of human organizations. In my view, it will be easier to meet this challenge if we understand in exactly what way organizations are alive.

Living social systems, as we have seen, are self-generating networks of communications.[23] This means that a human organization will be a living system only if it is organized as a network or contains smaller networks within its boundaries. Indeed, recently networks have become a major focus of attention not only in business but also in society at large and throughout a newly emerging global culture.

Within a few years, the Internet has become a powerful global network of communications, and many of the new Internet companies act as interfaces between networks of customers and suppliers. The pioneering example of this new type of organizational structure is Cisco Systems, a San Francisco company that is the largest provider of switches and routers for the Internet but that for many years did not own a single factory. Essentially, what Cisco does is produce and manage information through its web site by establishing contacts between suppliers and customers and by providing expert knowledge.[24]

Most large corporations today exist as decentralized networks of smaller units. In addition, they are connected to networks of small and medium businesses that serve as their subcontractors and suppliers, and units belonging to different corporations also enter into strategic alliances and engage in joint ventures. The various parts of those corporate networks continually recombine and interlink, cooperating and competing with one another at the same time.

Similar networks exist among nonprofit and nongovernmental organizations (NGOs). Teachers in schools and between schools increasingly interconnect through electronic networks, which also include parents and various organizations providing educational support. Moreover, networking has been one of the main activities of political grassroots organizations for many years. The environmental movement, the human rights movement, the feminist movement, the peace movement, and many other political and cultural grassroots movements have organized themselves as networks that transcend national boundaries.[25]

In 1999, hundreds of these grassroots organizations interlinked electronically for several months to prepare for joint protest actions at the meeting of the World Trade Organization (WTO) in Seattle. The Seattle Coalition was extremely successful in derailing the WTO meeting and in making its views known to the world. Its concerted actions, based on network strategies, have permanently changed the political climate around the issue of economic globalization.[26]

These recent developments make it evident that networks have become one of the most prominent social phenomena of our time. Social network analysis has become a new approach to sociology, and is employed by numerous scientists to study social relationships and the nature of community.[27] Turning to a larger scale, sociologist Manuel Castells argues that the recent information technology revolution has given rise to a new economy, structured around flows of information, power, and wealth in global financial networks. Castells also observes that throughout society, networking has emerged as a new form of organization of human activity, and he has coined the term "network society" to describe and analyze this new social structure.[28]

Communities of Practice

With the new information and communication technologies, social networks have become all-pervasive, both within and beyond organizations. For an organization to be alive, however, the existence of social networks is not sufficient; they need to be networks of a special type. Living networks, as we have seen, are self-generating. Each communication creates thoughts and meaning, which give rise to further communications. In this way, the entire network generates itself, producing a common context of meaning, shared knowledge, rules of conduct, a boundary, and a collective identity for its members.

Organizational theorist Etienne Wenger has coined the term "communities of practice" for these self-generating social networks, referring to the common context of meaning rather than to the pattern of organization through which the meaning is generated. "As people pursue any shared enterprise over time," Wenger explains, "they develop a common practice, that is, shared ways of doing things and relating to one another that allow them to achieve their joint purpose. Over time, the resulting practice becomes a recognizable bond among those involved."[29]

Wenger emphasizes that there are many different kinds of communities, just as there are many different kinds of social networks. A residential neighborhood, for example, is often called a community, and we also speak of the "legal community" or the "medical community." However, these are generally not communities of practice with the characteristic dynamics of self-generating networks of communications.

Wenger defines a community of practice as characterized by three features: mutual engagement of its members, a joint enterprise, and, over time, a shared repertoire of routines, tacit rules of conduct, and knowledge.[30] In terms of our conceptual framework, we see that the mutual engagement refers to the dynamics of a self-generating network of communications, the joint enterprise to the shared purpose and meaning, and the shared repertoire to the resulting coordination of behavior and creation of shared knowledge.

The generation of a common context of meaning, shared knowledge, and rules of conduct are characteristic of what I called the "dynamics of culture" in the preceding pages.[31] This includes, in particular, the creation of a boundary of meaning and hence of an identity among the members of the social network, based on a sense of belonging, which is the defining characteristic of community. According to Arie de Geus, a strong feeling among the employees of a company that they belong to the organization and identify with its achievements—in other words, a strong sense of community—is essential for the survival of companies in today's turbulent business environment.[32]

In our daily activities, most of us belong to several communities of practice—at work, in schools, in sports and hobbies, or in civic life. Some of them may have explicit names and formal structures, others may be so informal that they are not even identified as communities. Whatever their status, communities of practice are an integral part of our lives. As far as human organizations are concerned, we can now see that their dual nature as legal and economic entities, on the one hand, and communities of people on the other, derives from the fact that various communities of practice invariably arise and develop within the organization's formal structures. These are informal networks—alliances and friendships, informal channels of communication (the "grapevine"), and other tangled webs of relationships—that continually grow, change, and adapt to new situations. In the words of Etienne Wenger,

> Workers organize their lives with their immediate colleagues and customers to get their jobs done. In doing so, they develop or preserve a sense of themselves they can live with, have some fun, and fulfill the requirements of their employers and clients. No matter what their official job description may be, they create a practice to do what needs to be done. Although workers may be contractually employed by a large institution, in day-to-day practice they work with—and, in a sense, for—a much smaller set of people and communities.[33]

Within every organization, there is a cluster of interconnected communities of practice. The more people are engaged in these informal networks, and the more developed and sophisticated the networks are, the better will the organization be able to learn, respond creatively to unexpected new circumstances, change, and evolve. In other words, the organization's aliveness resides in its communities of practice.

The Living Organization

In order to maximize a company's creative potential and learning capabilities, it is crucial for managers and business leaders to understand the interplay between the organization's formal, designed structures and its informal, self-generating networks.[34] The formal structures are sets of rules and regulations that define relationships between people and tasks, and determine the distribution of power. Boundaries are established by contractual agreements that delineate well-defined subsystems (departments) and functions. The formal structures are depicted in the organization's official documents—the organizational charts, bylaws, manuals, and budgets that describe the organization's formal policies, strategies, and procedures.

The informal structures, by contrast, are fluid and fluctuating networks of communications.[35] These communications include nonverbal forms of mutual engagement in a joint enterprise through which skills are exchanged and shared tacit knowledge is generated. The shared practice creates flexible boundaries of meaning that are often unspoken. The distinction of belonging to a network may be as simple as being able to follow certain conversations or knowing the latest gossip.

Informal networks of communications are embodied in the people who engage in the common practice. When new people join, the entire network may reconfigure itself; when people leave, the network will change again, or may even break down. In the formal organization, by contrast, functions, and power relations are more important than people, persisting over the years while people come and go.

In every organization, there is a continuous interplay between its informal networks and its formal structures. Formal policies and procedures are always filtered and modified by the informal networks, which allow workers to use their creativity when faced with unexpected and novel situations. The power of this interplay becomes strikingly apparent when employees engage in a work-to-rule protest. By working strictly according to the official manuals and procedures, they seriously impair the organization's functioning. Ideally, the formal organization recognizes and supports its informal networks of relationships and incorporates their innovations into its structures.

To repeat, the aliveness of an organization—its flexibility, creative potential, and learning capability—resides in its informal communities of practice. The formal parts of the organization may be "alive" to varying degrees, depending on how closely they are in touch with their informal networks. Experienced managers know how to work with the informal organization. They will typically let the formal structures handle the routine work and rely on the informal organization to help with tasks that go beyond the usual routine. They may also communicate critical information to certain people, knowing that it will be passed around and discussed through the informal channels.

These considerations imply that the most effective way to enhance an organization's potential for creativity and learning, to keep it vibrant and alive, is to support and strengthen its communities of practice. The first step in this endeavor will be to provide the social space for informal communications to flourish. Some companies may create special coffee counters to encourage informal gatherings; others may use bulletin boards, the company newsletter, a special library, offsite retreats or online chat rooms for the same purpose. If widely publicized within the company so that support by management is evident, these measures will liberate people's energies, stimulate creativity, and set processes of change in motion.

Learning from Life

The more managers know about the detailed processes involved in self-generating social networks, the more effective they will be in working with the organization's communities of practice. Let us see, then, what kinds of lessons for management can be derived from the systemic understanding of life.[36]

A living network responds to disturbances with structural changes, and it chooses both which disturbances to notice and how to respond.[37] What people notice depends on who they are as individuals, and on the cultural characteristics of their communities of practice. A message will get through to them not only because of its volume or frequency, but because it is meaningful to them.

Mechanistically oriented managers tend to hold on to the belief that they can control the organization if they understand how all its parts fit together. Even the daily experience that people's behavior contradicts their expectations does not make them doubt their basic assumption. On the contrary, it compels them to investigate the mechanisms of management in greater detail in order to be able to control them.

We are dealing here with a crucial difference between a living system and a machine. A machine can be controlled; a living system, according to the systemic understanding of life, can only be disturbed. In other words, organizations cannot be controlled through direct interventions, but they can be influenced by giving impulses rather than instructions. To change the conventional style of management requires a shift of perception that is anything but easy, but it also brings great rewards. Working with the processes inherent in living systems means that we do not need to spend a lot of energy to move an organization. There is no need to push, pull, or bully it to make it change. Force or energy are not the issue; the issue is meaning. Meaningful disturbances will get the organization's attention and will trigger structural changes.

Giving meaningful impulses rather than precise instructions may sound far too vague to managers used to striving for efficiency and predictable results, but it is well known that intelligent, alert people rarely carry out instructions exactly to the letter. They always modify and reinterpret them, ignore some parts and add others of their own making. Sometimes, it may be merely a change of emphasis, but people always respond with new versions of the original instructions.

This is often interpreted as resistance, or even sabotage, but it can be interpreted quite differently. Living systems always choose what to notice and how to respond. When people modify instructions, they respond creatively to a disturbance, because this is the essence of being alive. In their creative responses, the living networks within the organization generate and communicate meaning, asserting their freedom to continually re-create themselves. Even a passive, or passive aggressive, response is a way for people to display their creativity. Strict compliance can only be achieved at the expense of robbing people of their vitality and turning them into listless, disaffected robots. This consideration is especially important in today's knowledge-based organizations, in which loyalty, intelligence, and creativity are the highest assets.

The new understanding of the resistance to mandated organizational change can be very powerful, as it allows us to work with people's creativity, rather than ignore it, and, indeed, to transform it into a positive force. If we involve people in the change process right from the start, they will "choose to be disturbed," because the process itself is meaningful to them. According to Wheatley and Kellner-Rogers:

> We have no choice but to invite people into the process of rethinking, redesigning, restructuring the organization. We ignore people's need to participate at our own peril. If they're involved, they will create a future that already has them in it. We won't have to engage in the impossible and exhausting tasks of "selling" them the solution, getting them "to enroll," or figuring out the incentives that might bribe them into compliant behaviours . . . In our experience, enormous struggles with implementation are created every time we *deliver* changes to the organization rather than figuring out how to involve people in their creation . . . [On the other hand,] we have seen implementation move with dramatic speed among people who have been engaged in the design of those changes.[38]

The task is to make the process of change meaningful to people right from the start, to get their participation, and to provide an environment in which their creativity can flourish.

Offering impulses and guiding principles rather than strict instructions evidently amounts to significant changes in power relations, from domination and control to cooperation and partnerships. This, too, is a fundamental implication of the new understanding of life. In recent years, biologists and ecologists have begun to shift their metaphors from hierarchies to networks and have come to realize that partnership—the tendency to associate, establish links, cooperate, and maintain symbiotic relationships—is one of the hallmarks of life.[39]

In terms of our previous discussion of power, we could say that the shift from domination to partnership corresponds to a shift from coercive power, which uses threats of sanctions to assure adherence to orders, and compensatory power, which offers financial incentives and rewards, to conditioned power, which tries to make instructions meaningful through persuasion and education.[40] Even in traditional organizations, the power embodied in the organization's formal structures is always filtered, modified, or subverted by communities of practice that create their own interpretations, as orders come down through the organizational hierarchy.

Organizational Learning

With the critical importance of information technology in today's business world, the concepts of knowledge management and organizational learning have become a central focus of management theory. The exact nature of organizational learning has been the subject of an ardent debate. Is a learning organization a social system capable of learning, or is it a community that encourages and supports the learning of its members? In other words, is learning only an individual or also a social phenomenon?

Organizational theorist Ilkka Tuomi reviews and analyzes recent contributions to this debate in a remarkable book, *Corporate Knowledge*, in which he proposes an integrative theory of knowledge management.[41] Tuomi's model of knowledge creation is based on earlier work by Ikujiro Nonaka, who introduced the concept of the "knowledge-creating company" into management theory and has been one of the main contributors to the new field of knowledge management.[42] Tuomi's views on organizational learning are very compatible with the ideas developed in the preceding pages. Indeed, I believe that the systemic understanding of reflective consciousness and social networks can contribute significantly to clarifying the dynamics of organizational learning.

According to Nonaka and his collaborator Hirotaka Takeuchi:

> In a strict sense, knowledge is created only by individuals . . . Organizational knowledge creation, therefore, should be understood as a process that "organizationally" amplifies the knowledge created by individuals and crystallizes it as a part of the knowledge network of the organization.[43]

At the core of Nonaka and Takeuchi's model of knowledge creation lies the distinction between explicit and tacit knowledge, which was introduced by philosopher Michael Polanyi in the 1980s. Whereas explicit knowledge can be communicated and documented through language, tacit knowledge is acquired through experience and often remains intangible. Nonaka and Takeuchi argue that, although knowledge is always created by individuals, it can be brought to light and expanded by the organization through social interactions in which tacit knowledge is transformed into explicit knowledge. Thus, while knowledge creation is an individual process, its amplification and expansion are social processes that take place *between* individuals.[44]

As Tuomi points out it is really impossible to separate knowledge neatly into two different "stocks." For Polanyi, tacit knowledge is always a precondition for explicit knowledge. It provides the context of meaning from which the knower acquires explicit knowledge. This unspoken context, also known as "common sense," which arises from a web of cultural conventions, is well-known to researchers in artificial intelligence as a major source of frustration. It is the reason why, after several decades of strenuous effort, they have still not succeeded in programming computers to understand human language in any significant sense.[45]

Tacit knowledge is created by the dynamics of culture resulting from a network of (verbal and nonverbal) communications within a community of practice. Organizational learning, therefore, is a social phenomenon, because the tacit knowledge on which all explicit knowledge is based is generated collectively. Moreover, cognitive scientists have come to realize that even the creation of explicit knowledge has a social dimension because of the intrinsically social nature of reflective consciousness.[46] The systemic understanding of life and cognition shows clearly that organizational learning has both individual and social aspects.

These insights have important implications for the field of knowledge management. They make it clear that the widespread tendency to treat knowledge as an entity that is independent of people and their social context—a thing that can be replicated, transferred, quantified, and traded—will not improve organizational learning. As Margaret Wheatley puts it, "If we want to succeed with knowledge management, we must attend to human needs and dynamics . . . Knowledge [is not] the asset or capital. People are."[47]

The systems view of organizational learning reinforces the lesson we have learned from the understanding of life in human organizations: the most effective way to enhance an organization's learning potential is to support and strengthen its communities of practice. In an organization that is alive, knowledge creation is natural and sharing what we have learned with friends and colleagues is humanly satisfying. To quote Wheatley once more: "Working for an organization that is intent on creating knowledge is a wonderful motivator, not because the organization will be more profitable, but because our lives will feel more worthwhile."[48]

The Emergence of Novelty

If the aliveness of an organization resides in its communities of practice, and if creativity, learning, change, and development are inherent in all living systems, how do these processes actually manifest in the organization's living networks and communities? To answer this question, we need to turn to a key characteristic of life that we have already encountered several times in the preceding pages—the spontaneous emergence of new order. The phenomenon of emergence takes place at critical points of instability that arise from fluctuations in the environment, amplified by feedback loops.[49] Emergence results in the creation of novelty that is often qualitatively different from the phenomena out of which it emerged. The constant generation of novelty—"nature's creative advance," as the philosopher Alfred North Whitehead called it—is a key property of all living systems.

In a human organization, the event triggering the process of emergence may be an offhand comment, which may not even seem important to the person who made it but is meaningful to some people in a community of practice. Because it is meaningful to them, they choose to be disturbed and circulate the information rapidly through the organization's networks. As it circulates through various feedback loops, the information may get amplified and expanded, even to such an extent that the organization can no longer absorb it in its present state. When that happens, a point of instability has been reached. The system cannot integrate the new information into its existing order; it is forced to abandon some of its structures, behaviors, or beliefs. The result is a state of chaos, confusion, uncertainty, and doubt; and out of that chaotic state a new form of order, organized around new meaning, emerges. The new order was not designed by any individual but emerged as a result of the organization's collective creativity.

This process involves several distinct stages. To begin with, there must be a certain openness within the organization, a willingness to be disturbed, in order to set the process in motion; and there has to be an active network of communications with multiple feedback loops to amplify the triggering event. The next stage is the point of instability, which may be experienced as tension, chaos, uncertainty, or crisis. At this stage, the system may either break *down*, or it may break *through* to a new state of order, which is characterized by novelty and involves an experience of creativity that often feels like magic.

Let us take a closer look at these stages. The initial openness to disturbances from the environment is a basic property of all life. Living organisms need to be open to a constant flow of resources (energy and matter) to stay alive; human organizations need to be open to a flow of mental resources (information and ideas), as well as to the flows of energy and materials that are part of the production of goods or services. The openness of an organization to new concepts, new technologies, and new knowledge is an indicator of its aliveness, flexibility, and learning capabilities.

The experience of the critical instability that leads to emergence usually involves strong emotions—fear, confusion, self-doubt, or pain—and may even amount to an existential crisis. This was the experience of the small community of quantum physicists in the 1920s, when their exploration of the atomic and subatomic world brought them into contact with a strange and unexpected reality. In their struggle to comprehend this new reality, the physicists became painfully aware that their basic concepts, their language, and their whole way of thinking were inadequate for describing atomic phenomena. For many of them, this period was an intense emotional crisis, as described most vividly by Werner Heisenberg:

> I remember discussions with Bohr which went through many hours till very late at night and ended almost in despair; and when at the end of the discussion I went alone for a walk in the neighbouring park I repeated to myself again and again the question: Can nature possibly be so absurd as it seemed to us in these atomic experiments?[50]

It took the quantum physicists a long time to overcome their crisis, but in the end the reward was great. From their intellectual and emotional struggles emerged deep insights into the nature of space, time, and matter, and with them the outlines of a new scientific paradigm.[51]

The experience of tension and crisis before the emergence of novelty is well known to artists, who often find the process of creation overwhelming and yet persevere in it with discipline and passion. Marcel Proust offers a beautiful testimony of the artist's experience in his masterpiece *In Search of Lost Time:*

> It is often simply from want of the creative spirit that we do not go to the full extent of suffering. And the most terrible reality brings us, with our suffering, the joy of a great discovery, because it merely gives a new and clear form to what we have long been ruminating without suspecting it.[52]

Not all experiences of crisis and emergence need to be that extreme, of course. They occur in a wide range of intensities, from small sudden insights to painful and exhilarating transformations. What they have in common is a sense of uncertainty and loss of control that is, at the very least, uncomfortable. Artists and other creative people know how to embrace this uncertainty and loss of control. Novelists often report how their characters take on lives of their own in the process of creation, as the story seems to write itself; and the great Michelangelo gave us the unforgettable image of the sculptor chipping away the excess marble to let the statue emerge.

After prolonged immersion in uncertainty, confusion, and doubt, the sudden emergence of novelty is easily experienced as a magical moment. Artists and scientists have often described these moments of awe and wonder when a confused and chaotic situation crystallizes miraculously to reveal a novel idea or a solution to a previously intractable problem. Since the process of emergence is thoroughly nonlinear, involving multiple feedback loops, it cannot be fully analyzed with our conventional, linear ways of reasoning, and hence we tend to experience it with a sense of mystery.

In human organizations, emergent solutions are created within the context of a particular organizational culture, and generally cannot be transferred to another organization with a different culture. This tends to be a big problem for business leaders who, naturally, are very keen on replicating successful organizational change. What they tend to do is replicate a new structure that has been successful without transferring the tacit knowledge and context of meaning from which the new structure emerged.

Emergence and Design

Throughout the living world, the creativity of life expresses itself through the process of emergence. The structures that are created in this process—the biological structures of living organisms as well as social structures in human communities—may appropriately be called "emergent structures." Before the evolution of humans, all living structures on the planet were emergent structures. With human evolution, language, conceptual thought, and all the other characteristics of reflective consciousness came into play. This enabled us to form mental images of physical objects, to formulate goals and strategies, and thus to create structures by design.

We sometimes speak of the structural "design" of a blade of grass or an insect's wing, but in doing so we use metaphorical language. These structures were not designed; rather, they were formed during the evolution of life and survived through natural selection. They are emergent structures. Design requires the ability to form mental images, and since this ability, as far as we know, is limited to humans and the other great apes, there is no design in nature at large.

Designed structures are always created for a purpose and embody some meaning.[53] In nonhuman nature, there is no purpose or intention. We often tend to attribute a purpose to the form of a plant or the behavior of an animal. For example, we would say that a flower has a certain color to attract honey bees, or that a squirrel hides its nuts in order to have a storage of food in winter, but these are anthropomorphic projections that ascribe the human characteristic of purposeful action to nonhuman phenomena. The colors of flowers and the behavior of animals have been shaped through long processes of evolution and natural selection, often in coevolution with other species. From the scientific point of view, there is neither purpose nor design in nature.[54]

This does not mean that life is purely random and meaningless, as the mechanistic neo-Darwinist school of thought would have it. The systemic understanding of life recognizes the pervasive order, self-organization, and intelligence manifest throughout the living world, and, as we have seen, this realization is completely consistent with a spiritual outlook on life.[55] However, the teleological assumption that purpose is inherent in natural phenomena is a human projection, because purpose is a characteristic of reflective consciousness, which does not exist in nature at large.[56]

Human organizations always contain both designed and emergent structures. The designed structures are the formal structures of the organization, as described in its official documents. The emergent structures are created by the organization's informal networks and communities of practice. The two types of structures are very different, as we have seen, and every organization needs both kinds.[57] Designed structures provide the rules and routines that are necessary for the effective functioning of the organization. They enable a business organization to optimize its production processes and to sell its products through effective marketing campaigns. Designed structures provide stability.

Emergent structures, on the other hand, provide novelty, creativity, and flexibility. They are adaptive, capable of changing and evolving. In today's complex business environment, purely designed structures do not have the necessary responsiveness and learning capability. They may be capable of magnificent feats, but since they are not adaptive, they are deficient when it comes to learning and changing, and thus liable to be left behind.

The issue is not one of discarding designed structures in favor of emergent ones. We need both. In every human organization there is a tension between its designed structures, which embody relationships of power, and its emergent structures, which represent the organization's aliveness and creativity. As Margaret Wheatley puts it, "The difficulties in organizations are manifestations of life asserting itself against the powers of control."[58] Skillful managers understand the interdependence between design and emergence. They know that in today's turbulent business environment, their challenge is to find the right balance between the creativity of emergence and the stability of design.

Two Kinds of Leadership

Finding the right balance between design and emergence seems to require the blending of two different kinds of leadership. The traditional idea of a leader is that of a person who is able to hold a vision, to articulate it clearly and to communicate it with passion and charisma. It is also a person whose actions embody certain values that serve as a standard for others to strive for. The ability to hold a clear vision of an ideal form, or state of affairs, is something that traditional leaders have in common with designers.

The other kind of leadership consists in facilitating the emergence of novelty. This means creating conditions rather than giving directions, and using the power of authority to empower others. Both kinds of leadership have to do with creativity. Being a leader means creating a vision; it means going where nobody has gone before. It also means enabling the community as a whole to create something new. Facilitating emergence means facilitating creativity.

Holding a vision is central to the success of any organization, because all human beings need to feel that their actions are meaningful and geared toward specific goals. At all levels of the organization, people need to have a sense of where they are going. A vision is a mental image of what we want to achieve, but visions are much more complex than concrete goals and tend to defy expression in ordinary, rational terms. Goals can be measured, while vision is qualitative and much more intangible.

Whenever we need to express complex and subtle images, we make use of metaphors, and thus it is not surprising that metaphors play a crucial role in formulating an organization's vision.[59] Often, the vision remains unclear as long as we try to explain it, but suddenly comes into focus when we find the right metaphor. The ability to express a vision in metaphors, to articulate it in such a way that it is understood and embraced by all, is an essential quality of leadership.

To facilitate emergence effectively, community leaders need to recognize and understand the different stages of this fundamental life process. As we have seen, emergence requires an active network of communications with multiple feedback loops. Facilitating emergence means first of all building up and nurturing networks of communications in order to "connect the system to more of itself," as Wheatley and Kellner-Rogers put it.[60]

In addition, we need to remember that the emergence of novelty is a property of open systems, which means that the organization needs to be open to new ideas and new knowledge. Facilitating emergence includes creating that openness—a learning culture in which continual questioning is encouraged and innovation is rewarded. Organizations with such a culture value diversity and, in the words of Arie de Geus, "tolerate activities in the margin: experiments and eccentricities that stretch their understanding."[61]

Leaders often find it difficult to establish the feedback loops that increase the organization's connectedness. They tend to turn to the same people again and again—usually the most powerful in the organization, who often resist change. Moreover, chief executives often feel that, because of the organization's traditions and past history, certain delicate issues cannot be addressed openly.

In those cases, one of the most effective approaches for a leader may be to hire an outside consultant as a "catalyst." Being a catalyst means that the consultant is not affected by the processes she helps to initiate, and thus is able to analyze the situation much more clearly. Angelika Siegmund, cofounder of Corphis Consulting in Munich, Germany, describes this work in the following words:

> One of my main activities is to act as feedback facilitator and amplifier. I don't design solutions but facilitate feedback; the organization takes care of the contents. I analyse the situation, reflect it back to management, and make sure that every decision is immediately communicated through a feedback loop. I build up networks, increase the organization's connectivity, and amplify the voices of employees who would otherwise not be heard. As a consequence, the managers begin to discuss things that would normally not be discussed, and thus the organization's ability to learn increases. In my experience, a powerful leader plus a skilled outside facilitator is a fantastic combination that can bring about incredible effects.[62]

The experience of the critical instability that precedes the emergence of novelty may involve uncertainty, fear, confusion, or self-doubt. Experienced leaders recognize these emotions as integral parts of the whole dynamic and create a climate of trust and mutual support. In today's turbulent global economy this is especially important, because people are often in fear of losing their jobs as a consequence of corporate mergers or other radical structural changes. This fear generates a strong resistance to change, hence building trust is essential.

The problem is that people at all levels want to be told what concrete results they can expect from the change process, while managers themselves do not know what will emerge. During this chaotic phase, many managers tend to hold things back rather than communicating honestly and openly, which means that rumors fly and nobody knows what information to trust.

Good leaders will tell their employees openly and often which aspects of the change have been established and which are still uncertain. They will try to make the process transparent, even though the results cannot be known in advance.

During the change process some of the old structures may fall apart, but if the supportive climate and the feedback loops in the network of communications persist, new and more meaningful structures are likely to emerge. When that happens, people often feel a sense of wonder and elation, and now the leader's role is to acknowledge these emotions and provide opportunities for celebration.

Finally, leaders need to be able to recognize emergent novelty, articulate it and incorporate it into the organization's design. Not all emergent solutions will be viable, however, and hence a culture fostering emergence must include the freedom to make mistakes. In such a culture, experimentation is encouraged and learning is valued as much as success.

Since power is embodied in all social structures, the emergence of new structures will always change power relations; the process of emergence in communities is also a process of collective empowerment. Leaders who facilitate emergence use their own power to empower others. The result may be an organization in which both power and the potential for leadership are widely distributed. This does not mean that several individuals assume leadership simultaneously, but that different leaders step forward when they are needed to facilitate various stages of emergence. Experience has shown that it usually takes years to develop this kind of distributed leadership.

It is sometimes argued that the need for coherent decisions and strategies requires an ultimate seat of power. However, many business leaders have pointed out that coherent strategy emerges when senior executives are engaged in an ongoing process of conversation. In the words of Arie de Geus, "Decisions grow in the topsoil of formal and informal conversation—sometimes structured (as in board meetings and the budget process), sometimes technical (devoted to implementation of specific plans or practices), and sometimes ad hoc."[63]

Different situations will require different types of leadership. Sometimes, informal networks and feedback loops will have to be established; at other times people will need firm frameworks with definite goals and time frames within which they can organize themselves. An experienced leader will assess the situation, take command if necessary, but then be flexible enough to let go again. It is evident that such leadership requires a wide variety of skills, so that many paths for action are available.

Bringing Life into Organizations

Bringing life into human organizations by empowering their communities of practice not only increases their flexibility, creativity, and learning potential, but also enhances the dignity and humanity of the organization's individuals, as they connect with those qualities in themselves. In other words, the focus on life and self-organization empowers the self. It creates mentally and emotionally healthy working environments in which people feel that they are supported in striving to achieve their own goals and do not have to sacrifice their integrity to meet the goals of the organization.

The problem is that human organizations are not only living communities but are also social institutions designed for specific purposes and functioning in a specific economic environment. Today that environment is not life-enhancing but is increasingly life-destroying. The more we understand the nature of life and become aware of how alive an organization can be, the more painfully we notice the life-draining nature of our current economic system.

When shareholders and other outside bodies assess the health of a business organization, they generally do not inquire about the aliveness of its communities, the integrity and well-being of its employees, or the ecological sustainability of its products. They ask about profits, shareholder value, market share, and other economic parameters; and they will apply any pressure they can to assure quick returns on their investments, irrespective of the long-term consequences for the organization, the well-being of its employees, or of its broader social and environmental impacts.

These economic pressures are applied with the help of ever more sophisticated information and communication technologies, which have created a profound conflict between biological time and computer time. New knowledge arises, as we have seen, from chaotic processes of emergence that take time. Being creative means being able to relax into uncertainty and confusion. In most organizations this is becoming increasingly difficult, because things move far too fast. People feel that they have hardly any time for quiet reflection, and since reflective consciousness is one of the defining characteristics of human nature, the results are profoundly dehumanizing.

The enormous workload of today's executives is another direct consequence of the conflict between biological time and computer time. Their work is increasingly computerized, and as computer technology progresses, these machines work faster and faster and thus save more and more time. What to do with that spare time becomes a question of values. It can be distributed among the individuals in the organization—thus creating time for them to reflect, organize themselves, network, and gather for informal conversations—or the time can be extracted from the organization and turned into profits for its top executives and shareholders by making people work more and thus increasing the company's productivity. Unfortunately, most companies in our much-acclaimed information age have chosen the second option. As a consequence, we see enormous increases in the corporate wealth at the top, while thousands of workers are fired in the continuing mania for downsizing and corporate mergers, and those remaining (including the top executives themselves) are forced to work harder and harder.

Most corporate mergers involve dramatic and rapid structural changes for which people are totally unprepared. Acquisitions and mergers are undertaken partly because large corporations want to gain entry into new markets and buy knowledge or technologies developed by smaller companies (in the mistaken belief that they can short-circuit the learning process). Increasingly, however, the main reason for a merger is to make the company bigger and thus less susceptible to being swallowed itself. In most cases, a merger involves a highly problematic fusion of two different corporate cultures, which seems to bring no advantages in terms of greater efficiency or profits, but produces protracted power struggles, enormous stress, existential fears, and thus deep distrust and suspicions about structural change.[64]

It is evident that the key characteristics of today's business environment—global competition, turbulent markets, corporate mergers with rapid structural changes, increasing workloads, and demands for "24/7" accessibility through e-mail and cell phones—combine to create a situation that is highly stressful and profoundly unhealthy. In this climate it is often difficult to hold on to the vision of an organization that is alive, creative, and concerned about the well-being of its members and of the living world at large. When we are under stress, we tend to revert to old ways of acting. When things fall apart in a chaotic situation, we tend to take hold and assume control. This tendency is especially strong among managers, who are used to getting things done and are attracted to the exercise of control.

Paradoxically, the current business environment, with its turbulences and complexities and its emphasis on knowledge and learning, is also one in which the flexibility, creativity, and learning capability that come with the organization's aliveness are most needed. This is now being recognized by a growing number of visionary business leaders who are shifting their priorities toward developing the creative potential of their employees, enhancing the quality of the company's internal communities, and integrating the challenges of ecological sustainability into their strategies. Because of the need for continuous change management in today's turbulent environment, the "learning organizations" managed by this new generation of business leaders are often very successful in spite of present economic constraints.[65]

In the long run, organizations that are truly alive will be able to flourish only when we change our economic system so that it becomes life-enhancing rather than life-destroying. This is a global issue, which I shall discuss in some detail in the following pages. We shall see that the life-draining characteristics of the economic environment in which today's organizations have to operate are not isolated, but are invariably consequences of the "new economy" that has become the critical context of our social and organizational life.

This new economy is structured around flows of information, power, and wealth in global financial networks that rely decisively on advanced information and communication technologies.[66] It is shaped in very fundamental ways by machines, and the resulting economic, social, and cultural environment is not life-enhancing but life-degrading. It has triggered a great deal of resistance, which may well coalesce into a worldwide movement to change the current economic system by organizing its financial flows according to a different set of values and beliefs. The systemic understanding of life makes it clear that in the coming years such a change will be imperative not only for the well-being of human organizations, but also for the survival and sustainability of humanity as a whole.

| five |

THE NETWORKS OF GLOBAL CAPITALISM

During the last decade of the twentieth century, a recognition grew among entrepreneurs, politicians, social scientists, community leaders, grassroots activists, artists, cultural historians, and ordinary women and men from all walks of life that a new world was emerging—a world shaped by new technologies, new social structures, a new economy and a new culture. "Globalization" became the term used to summarize the extraordinary changes and the seemingly irresistible momentum felt by millions of people.

With the creation of the World Trade Organization (WTO) in the mid-1990s, economic globalization, characterized by "free trade," was hailed by corporate leaders and politicians as a new order that would benefit all nations, producing worldwide economic expansion whose wealth would trickle down to all. However, it soon became apparent to increasing numbers of environmentalists and grassroots activists that the new economic rules established by the WTO were manifestly unsustainable and were producing a multitude of interconnected fatal consequences—social disintegration, a breakdown of democracy, more rapid and extensive deterioration of the environment, the spread of new diseases, and increasing poverty and alienation.

Understanding Globalization

In 1996, two books were published that provided the first systemic analyses of the new economic globalization. They are written in very different styles and their authors follow very different approaches, but their starting point is the same—the attempt to understand the profound changes brought about by the combination of extraordinary technological innovation and global corporate reach.

The Case Against the Global Economy is a collection of essays by more than forty grassroots activists and community leaders, edited by Jerry Mander and Edward Goldsmith, and published by the Sierra Club, one of the oldest and most respected environmental organizations in the United States.[1] The authors of this book represent cultural traditions from many countries around the world. Most of them are well known among social-change activists. Their arguments are passionate, distilled from the experiences of their communities, and aimed at reshaping globalization according to different values and different visions.

The Rise of the Network Society by Manuel Castells, Professor of Sociology at the University of California at Berkeley, is a brilliant analysis of the fundamental processes underlying economic globalization, published by Blackwell, one of the largest academic publishers.[2] Castells believes that, before attempting to reshape globalization, we need to understand the deep systemic roots of the world that is now emerging. "I propose the hypothesis," he writes in the prologue to his book, "that all major trends of change constituting our new, confusing world are related, and that we can make sense of their interrelationship. And, yes, I believe, in spite of a long tradition of sometimes tragic intellectual errors, that observing, analysing, and theorizing is a way of helping to build a different, better world."[3]

During the years following the publication of these two books, some of the authors of *The Case Against the Global Economy* formed the International Forum on Globalization, a nonprofit organization that holds teach-ins on economic globalization in several countries. In 1999, these teach-ins provided the philosophical background for the worldwide coalition of grassroots organizations that successfully blocked the meeting of the World Trade Organization in Seattle and made its opposition to the WTO's policies and autocratic regime known to the world.

On the theoretical front, Manuel Castells published two further books, *The Power of Identity* (1997) and *End of Millennium* (1998) to complete a series of three volumes on *The Information Age: Economy, Society and Culture.*[4] This trilogy is a monumental work, encyclopedic in its rich documentation, which Anthony Giddens has compared to Max Weber's *Economy and Society*, written almost a century earlier.[5]

Castells's thesis is wide-ranging and illuminating. His central focus is on the revolutionary information and communication technologies that emerged during the last three decades of the twentieth century. As the Industrial Revolution gave rise to "industrial society," so the new Information Technology Revolution is now giving rise to an "informational society." And since information technology has played a decisive role in the rise of networking as a new form of organization of human activity in business, politics, the media and in nongovernmental organizations, Castells also calls the informational society the "network society."

Another important and rather mysterious aspect of globalization was the sudden collapse of Soviet communism in the 1980s, which occurred without the intervention of social movements and without a major war, and which came as a complete surprise to most Western observers. According to Castells, this profound geopolitical transformation, too, was a consequence of the Information Technology Revolution. In a detailed analysis of the economic demise of the Soviet Union, Castells postulates that the roots of the crisis that triggered Gorbachev's *perestroika* and eventually led to the breakup of the USSR are found in the inability of the Soviet economic and political system to navigate the transition to the new informational paradigm that was spreading through the rest of the world.[6]

Since the demise of Soviet communism, capitalism has been thriving throughout the world and, as Castells observes, "it deepens its penetration of countries, cultures, and domains of life. In spite of a highly diversified social and cultural landscape, for the first time in history, the whole world is organized around a largely common set of economic rules."[7]

During the first years of this new century, the attempts of scholars, politicians, and community leaders to understand the nature and consequences of globalization have continued and intensified. In 2000, a collection of essays on global capitalism by some of the world's leading political and economic thinkers was published by British social scientists Will Hutton and Anthony Giddens.[8] At the same time, Czech president Václav Havel and Nobel laureate Elie Wiesel assembled a distinguished group of religious leaders, politicians, scientists and community leaders in a series of annual symposia, called "Forum 2000," at Prague Castle to engage in discussions "about the problems of our civilization . . . [and to] think about the political dimension, the human dimension, and the ethical dimension of globalization."[9]

In this chapter, I shall try to synthesize the main ideas about globalization that I have learned from the people and publications mentioned above. In doing so, I hope to contribute some insights of my own from the perspective of the new unified understanding of biological and social life that I presented in the first three chapters of this book. In particular, I shall try to show how the rise of globalization has proceeded through a process that is characteristic of all human organizations—the interplay between designed and emergent structures.[10]

The Information Technology Revolution

The common characteristic of the multiple aspects of globalization is a global information and communications network based on revolutionary new technologies. The Information Technology Revolution is the result of a complex dynamic of technological and human interactions, which produced synergistic effects in three major areas of electronics—computers, microelectronics, and telecommunications. The key innovations that created the radically new electronic environment of the 1990s all took place twenty years earlier, during the 1970s.[11]

Computer technology is based theoretically on cybernetics, which is also one of the conceptual roots of the new systemic understanding of life.[12] The first commercial computers were produced in the 1950s, and during the 1960s IBM established itself as the dominant force in the computer industry with its large mainframe machines. The development of microelectronics during the following years changed this picture dramatically. It began with the invention and subsequent miniaturization of the integrated circuit—a tiny electronic circuit embedded in a "chip" of silicon—which may contain thousands of transistors that process electric impulses.

In the early 1970s, microelectronics took a giant leap with the invention of the microprocessor, which is essentially a computer on a chip. Since then, the density (or "integration capacity") of circuits on these microprocessors has increased phenomenally. In the 1970s, thousands of transistors were packed on a chip the size of a thumbnail; twenty years later, it was millions. Computing capacity increased relentlessly with the advance of microelectronics into dimensions so small that they defy imagination. And as these information-processing chips became smaller and smaller they were placed in virtually all the machines and appliances of our everyday life, where we are not even aware of their existence.

The application of microelectronics to computer design led to a dramatic reduction in computer size within a few years. The launch of the first Apple microcomputer in the mid-seventies by two young college dropouts, Steve Jobs and Stephen Wozniak, shattered the dominance of the old mainframes. But IBM was quick to respond by introducing its own microcomputer with the ingenious name "the Personal Computer (PC)," which soon became the generic name for microcomputers.

In the mid-eighties Apple launched its first Macintosh, featuring the user-friendly icon-and-mouse technology. At the same time, another pair of young college dropouts, Bill Gates and Paul Allen, created the first PC software and, based on this success, founded Microsoft, today's software giant.

The current stage of the Information Technology Revolution was reached when the advanced PC technologies and microelectronics were combined synergistically with the latest achievements in telecommunication. The worldwide communications revolution had begun in the late 1960s when the first satellites were put into stationary orbits and used to transmit signals between any two points on the Earth almost instantly. Today's satellites can handle thousands of communication channels simultaneously. Some of them also provide a constant signal that allows aircraft, ships, and even individual cars to determine their positions with great accuracy.

In the meantime, surface communications on Earth intensified, with major advances in fiber optics that dramatically increased the capacity of transmission lines. Whereas the first transatlantic telephone cable in 1956 carried fifty compressed voice channels, today's optical-fiber cables carry over 50,000. In addition, the diversity and versatility of communications increased considerably through the use of a greater variety of electromagnetic frequencies, including those of microwaves, laser transmission, and digital cellular telephones.

The combined effect of all these developments on the use of computers has been a dramatic shift from data storage and processing in large, isolated machines to the interactive use of microcomputers and the sharing of computer power in electronic networks. The outstanding example of this new form of interactive computer use is, of course, the Internet, which grew in less than three decades from a small experimental network, serving a dozen research institutes in the United States, to a global system of thousands of interconnected networks, linking millions of computers, and capable of seemingly infinite expansion and diversification. The evolution of the Internet is a fascinating story. It exemplifies in the most dramatic way the continual interplay between ingenious design and spontaneous emergence that has been characteristic of the Information Technology Revolution as a whole.[13]

In Europe and the United States, the 1960s and 1970s were not only a time of revolutionary technological innovations but also one of social upheavals. From the Civil Rights movement in the American South to the Free Speech movement on the Berkeley campus, the Prague Spring and the "May '68" student revolt in Paris, a worldwide counterculture emerged that championed the questioning of authority, a sense of personal freedom and empowerment, and the expansion of consciousness, both spiritually and socially. The artistic expressions of these ideals generated many new styles and movements in the arts, producing powerful new forms of poetry, theater, film, music, and dance that defined the zeitgeist of that period.

The social and cultural innovations of the sixties and seventies not only shaped the subsequent decades in many ways, but also influenced some of the leading innovators in the Information Technology Revolution. When Silicon Valley became the new technological frontier and attracted thousands of creative young minds from around the world, these new pioneers soon discovered—if they did not know it already—that the San Francisco Bay Area was also a thriving center of the counterculture. The irreverent attitudes, strong sense of community and cosmopolitan sophistication of the sixties formed the cultural background of the informal, open, decentralized, cooperative, and future-oriented working styles that became characteristic of the new information technologies.[14]

The Rise of Global Capitalism

For several decades after World War II, the Keynesian model of capitalist economics, based on a social contract between capital and labor and on fine tuning the business cycles of national economies by centralized measures—raising or lowering interest rates, cutting or increasing taxes, etc.—was remarkably successful, bringing economic prosperity and social stability to most countries with mixed market economies. In the 1970s, however, the model reached its conceptual limitations.[15]

Keynesian economists concentrated on the domestic economy, disregarding international economic agreements and a growing global economic network; they neglected the overwhelming power of transnational corporations, which had become major actors on the global stage; and, last but not least, they ignored the social and environmental costs of economic activities, as most economists still do. When an oil crisis hit the industrialized world in the late 1970s, together with rampant inflation and massive unemployment, the impasse that Keynesian economics had reached became evident.

In response to the crisis, Western governments and business organizations engaged in a painful process of capitalist restructuring, while a parallel (but ultimately unsuccessful) process of communist restructuring—Gorbachev's *perestroika*—took place in the Soviet Union. The capitalist restructuring process involved the gradual dismantling of the social contract between capital and labor, the deregulation and liberalization of financial trading, and many organizational changes designed to increase flexibility and adaptability.[16] It proceeded pragmatically by trial and error and had very different impacts on different countries around the world—from the disastrous effects of "Reaganomics" in the United States and the resistance to the dismantling of the welfare state in Western Europe to the successful mix of high technology, competitiveness, and cooperation in Japan. Eventually, the capitalist restructuring imposed a common economic discipline on the countries of the emerging global economy, enforced by the central banks and the International Monetary Fund (IMF).

All these measures relied crucially on the new information and communication technologies, which made it possible to transfer funds between various segments of the economy and various countries almost instantly and to manage the enormous complexity brought about by rapid deregulation and new financial ingenuity. In the end, the Information Technology Revolution helped to give birth to a new global economy—a rejuvenated, flexible and greatly expanded capitalism.

As Castells emphasizes, this new capitalism is profoundly different from the one formed during the Industrial Revolution, or the one that emerged after World War II. It is characterized by three fundamental features; its core economic activities are global; the main sources of productivity and competitiveness are innovation, knowledge generation, and information processing; and it is structured largely around networks of financial flows.

The New Economy

In the new economy, capital works in real time, moving rapidly through global financial networks. From these networks it is invested in all kinds of economic activity, and most of what is extracted as profit is channelled back into the metanetwork of financial flows. Sophisticated information and communication technologies enable financial capital to move rapidly from one option to another in a relentless global search for investment opportunities. Profit margins are generally much higher in the financial markets than in most direct investments, hence, all flows of money ultimately converge in the global financial networks in search of higher gains.

The dual role of computers as tools for rapid processing of information and for sophisticated mathematical modelling has led to the virtual replacement of gold and paper money by ever more abstract financial products. These include "future options" (options to buy at a specific point in the future with the hope of reaping financial gains anticipated by computer projections), "hedge funds" (investment funds that are often used to buy and sell huge amounts of currencies within minutes to profit from tiny margins), and "derivatives" (packages of diverse funds, representing collections of actual or potential financial values). Here is how Manuel Castells describes the resulting global casino:

> The same capital is shuttled back and forth between economies in a matter of hours, minutes, and sometimes seconds. Favoured by deregulation . . . and the opening of domestic financial markets, powerful computer programs and skillful financial analysts/computer wizards sitting at the global nodes of a selective telecommunications network play games, literally, with billions of dollars . . . These global gamblers are not obscure speculators, but major investment banks, pension funds, multinational corporations . . . and mutual funds organized precisely for the sake of financial manipulation.[17]

With the increasing "virtuality" of financial products and the growing importance of computer models that are based on the subjective perceptions of their creators, the attention of investors has shifted from real profits to the subjective and volatile criterion of perceived stock value. In the new economy, the basic objective of the game is not so much to maximize profits as to maximize shareholder value. In the long run, of course, the value of a company will decrease if it keeps operating without making any profit, but in the short run its value may increase or decrease regardless of actual performance, based on often intangible market expectations.

The new Internet companies, or "dot-coms," which for a time showed skyrocketing increases in value without making profits, are striking examples of the decoupling of money-making from profit-making in the new economy. On the other hand, stock values of sound companies have also crashed dramatically, wrecking the companies and leading to massive job cuts in spite of continuing solid performance, merely because of subtle changes in the companies' financial environment.

To be competitive in the global network of financial flows, the rapid processing of information and the knowledge required for technological innovation are crucial. In the words of Castells: "Productivity essentially stems from innovation, competitiveness from flexibility . . . Information technology, and the cultural capacity to use it, are essential [for both]."[18]

Complexity and Turbulence

The process of economic globalization was purposefully designed by the leading capitalist countries (the so-called "G-7 nations"), the major transnational corporations, and by global financial institutions—most importantly, the World Bank, the International Monetary Fund (IMF) and the World Trade Organization (WTO)—that were created for that purpose.

However, the process has been far from smooth. Once the global financial networks reached a certain level of complexity, their nonlinear interconnections generated rapid feedback loops that gave rise to many unsuspected emergent phenomena. The resulting new economy is so complex and turbulent that it defies analysis in conventional economic terms. Thus Anthony Giddens, now the director of the prestigious London School of Economics, admits: "The new capitalism that is one of the driving forces of globalization to some extent is a mystery. We don't fully know as yet just how it works."[19]

In the electronically operated global casino, the financial flows do not follow any market logic. The markets are continually manipulated and transformed by computer-enacted investment strategies, subjective perceptions of influential analysts, political events in any part of the world, and—most significantly—by unsuspected turbulences caused by the complex interactions of capital flows in this highly nonlinear system. These largely uncontrolled turbulences are as important in setting prices and market trends as are the traditional forces of supply and demand.[20]

Global currency markets alone involve the daily exchange of over two trillion dollars, and since these markets largely determine the value of any national currency, they contribute significantly to the inability of governments to control economic policy.[21] As a result, we have seen a series of severe financial crises in recent years, from Mexico (1994) to the Asian Pacific (1997), Russia (1998), and Brazil (1999).

Large economies with strong banks are usually able to absorb financial turbulences with limited and temporary damage, but the situation is much more critical for the so-called "emerging markets" of the South, whose economies are tiny in comparison with international markets.[22] Because of their strong potential for economic growth, these countries have become prime targets for speculators in the global casino, who invest massively in emerging markets, but will remove their investments immediately at the first sign of weakness.

By doing so, they destabilize a small economy, induce capital flight, and create a full-blown crisis. To regain the confidence of investors, the afflicted country will typically be required by the IMF to raise its interest rates at the devastating cost of deepening the local recession.

The recent crashes of the financial markets threw approximately 40 percent of the world's population into deep recession![23]

In the wake of the Asian financial crisis, economists blamed a number of "structural factors" in Asian countries, including weak banking systems, government interference and lack of financial transparency. However, as Paul Volcker, the former Chair of the Federal Reserve Board, points out, none of these factors were new or unknown, nor had they suddenly become worse. "Quite obviously," Volcker concludes, "something has been lacking in our analyses and in our response . . . The problem is not regional, but international. And there is every indication that it is systemic."[24] According to Manuel Castells, the global financial networks of the new economy are inherently unstable. They produce random patterns of informational turbulence that may destabilize any company, as well as entire countries or regions, regardless of their economic performances.[25]

It is interesting to apply the systemic understanding of life to the analysis of this phenomenon. The new economy consists of a global metanetwork of complex technological and human interactions, involving multiple feedback loops operating far from equilibrium, which produce a never-ending variety of emergent phenomena. Its creativity, adaptability, and cognitive capabilities are certainly reminiscent of living networks, but it does not display the stability that is also a key property of life. The information circuits of the global economy operate at such speed and use such a multitude of sources that they constantly react to a flurry of information, and thus the system as a whole is spinning out of control.

Living organisms and ecosystems, too, may become continually unstable, but if they do, they will eventually disappear because of natural selection, and only those systems that have stabilizing processes built into them will survive. In the human realm, these processes will have to be introduced into the global economy through human consciousness, culture, and politics. In other words, we need to design and implement regulatory mechanisms to stabilize the new economy. As Robert Kuttner, editor of the progressive magazine *The American Prospect*, sums up the situation, "The stakes are simply too high to let speculative capital and currency swings determine the fate of the real economy."[26]

The Global Market—An Automaton

At the existential human level, the most alarming feature of the new economy may be that it is shaped in very fundamental ways by machines. The so-called "global market," strictly speaking, is not a market at all but a network of machines programmed according to a single value—money-making for the sake of making money—to the exclusion of all other values. In the words of Manuel Castells:

> The outcome of [the] process of financial globalization may be that we have created an Automaton at the core of our economies [that is] decisively conditioning our lives. Humankind's nightmare of seeing our machines taking control of our world seems on the edge of becoming reality—not in the form of robots that eliminate jobs or government computers that police our lives, but as an electronically based system of financial transactions.[27]

The logic of this automaton is not that of traditional market rules, and the dynamics of the financial flows it sets in motion is currently beyond the control of governments, corporations, and financial institutions, regardless of their wealth and power. However, because of the great versatility and accuracy of the new information and communication technologies, effective regulation of the global economy is technically feasible. The critical issue is not technology, but politics and human values.[28] And these human values can change; they are not natural laws. The same electronic networks of financial and informational flows *could* have other values built into them.

One important consequence of the exclusive focus on profits and shareholder value in the new global capitalism has been the mania for corporate mergers and acquisitions. In the global electronic casino, any share that can be sold for a higher profit *will* be sold, and this becomes the basis of the standard scenario for hostile takeovers. When a corporation wants to buy another company, all it has to do is offer a higher price for the company's shares. The legion of brokers whose job it is to scan the market constantly for investment and profit opportunities will then contact the shareholders and urge them to sell their shares for the higher price.

Once these hostile takeovers became possible, the owners of large corporations used them to gain entry into new markets, to buy special technologies developed by small companies or simply to grow and gain corporate prestige. The small companies, on the other hand, became afraid of being swallowed, and to protect themselves they bought still smaller ones in order to become larger and less easy to buy. Thus merger mania was unleashed, and there seems to be no end to it. Most corporate mergers, as mentioned above, seem to bring no advantages in terms of greater efficiency or profits, but do involve dramatic and rapid structural changes for which people are totally unprepared, and thus bring enormous stress and hardship.[29]

The Social Impact

In his trilogy on the Information Age, Manuel Castells provides a detailed analysis of the social and cultural impact of global capitalism. He describes in particular how the new network economy has profoundly transformed the social relationships between capital and labor. Money has become almost entirely independent of production and services by escaping into the virtual reality of electronic networks. Capital is global, while labor, as a rule, is local. Thus, capital and labor increasingly exist in different spaces and times: the virtual space of financial flows and the real space of the local and regional places where people are employed; the instant time of electronic communications and the biological time of everyday life.[30]

Economic power resides in the global financial networks, which determine the fate of most jobs, while labor remains locally constrained in the real world. Thus labor has become fragmented and disempowered. Many workers today, whether unionized or not, will not fight for higher wages or better working conditions out of fear that their jobs will be moved abroad.

As more and more companies restructure themselves as decentralized networks—networks of smaller units which, in turn, are linked to networks of suppliers and subcontractors—workers are employed increasingly through individual contracts, and labor is losing its collective identity and bargaining power. Indeed, in the new economy traditional working-class communities have all but disappeared.

Castells points out that it is important to distinguish between two kinds of labor. Unskilled, generic labor is not required to access information and knowledge beyond the ability to understand and execute orders. In the new economy, masses of generic workers move in and out of a variety of jobs. They may be replaced at any moment, either by machines or by generic labor in other parts of the world, depending on the fluctuations in the global financial networks.

"Self-educated" labor, by contrast, has the capacity to access higher levels of education, to process information, and to create knowledge. In an economy where information processing, innovation, and knowledge creation are the main sources of productivity, these self-educated workers are highly valued. Companies would like to maintain long-term, secure relationships with their core workers, so as to retain their loyalty and make sure that their tacit knowledge is passed on within the organization.

As an incentive to stay on, such workers are increasingly offered stock options in addition to their basic salaries, which gives them a stake in the value created by the company. This has further undermined the traditional class solidarity of labor. "The struggle between diverse capitalists and miscellaneous working classes," notes Castells, "is subsumed into the more fundamental opposition between the bare logic of capital flows and the cultural values of human experience."[31]

The new economy has certainly enriched a global elite of financial speculators, entrepreneurs, and high-tech professionals. At the very top, there has been an unprecedented accumulation of wealth, and global capitalism has also benefited some national economies, especially in Asian countries. Overall, however, its social and economic impacts have been disastrous.

The fragmentation and individualization of labor and the gradual dismantling of the welfare state under the pressures of economic globalization means that the rise of global capitalism has been accompanied by rising social inequality and polarization.[32] The gap between the rich and the poor has grown significantly, both internationally and within countries. According to the United Nation's Human Development Report, the difference in per capita income between the North and South tripled from $5,700 in 1960 to $15,000 in 1993. The richest 20 percent of the world's people now own 85 percent of its wealth, while the poorest 20 percent (who account for 80 percent of the total world population) owns just 1.4 percent.[33] The assets of the three richest people in the world alone exceed the combined GNP of all least developed countries and their 600 million people.[34]

In the United States, the wealthiest and technologically most advanced country in the world, median family income stagnated during the last three decades, and in California it even declined during the 1990s in the midst of the high-tech boom: most families today can make ends meet only if two members are contributing to the household budget.[35] The increase of poverty, and especially of extreme poverty, seems to be a worldwide phenomenon. Even in the United States, 15 percent of the population (including 25 percent of all children) now lives below the poverty line.[36] One of the most striking features of the "new poverty" is homelessness, which skyrocketed in American cities during the 1980s and remains at high levels today.

Global capitalism has increased poverty and social inequality not only by transforming the relationships between capital and labor, but also through the process of "social exclusion," which is a direct consequence of the new economy's network structure. As the flows of capital and information interlink worldwide networks, they exclude from these networks all populations and territories that are of no value or interest to their search for financial gain. As a result, certain segments of societies, areas of cities, regions, and even entire countries become economically irrelevant. In the words of Castells:

> Areas that are non-valuable from the perspective of informational capitalism, and that do not have significant political interest for the powers that be, are bypassed by flows of wealth and information, and ultimately deprived of the basic technological infrastructure that allows us to communicate, innovate, produce, consume, and even live, in today's world.[37]

The process of social exclusion is epitomized by the desolation of American inner-city ghettos, but its effects reach far beyond individuals, neighborhoods, and social groups. Around the world, a new impoverished segment of humanity has emerged that is sometimes referred to as the Fourth World. It comprises large areas of the globe, including much of Sub-Saharan Africa and impoverished rural areas of Asia and Latin America. The new geography of social exclusion includes portions of every country and every city in the world.[38]

The Fourth World is populated by millions of homeless, impoverished, and often illiterate people who move in and out of paid work, many of them drifting into the criminal economy. They experience multiple crises in their lives, including hunger, disease, drug addiction, and imprisonment—the ultimate form of social exclusion. Once their poverty turns into misery, they may easily find themselves caught in a downward spiral of marginality from which it is almost impossible to escape. Manuel Castells's detailed analysis of these disastrous social consequences of the new economy illuminates their systemic interconnections and adds up to a devastating critique of global capitalism.

The Ecological Impact

According to the doctrine of economic globalization—known as "neoliberalism," or "the Washington consensus"—the free-trade agreements imposed by the WTO on its member countries will increase global trade; this will create a global economic expansion; and global economic growth will decrease poverty, because its benefits will eventually "trickle down" to all. As political and corporate leaders like to say, the rising tide of the new economy will lift all boats.

Castells's analysis shows clearly that this reasoning is fundamentally flawed. Global capitalism does not alleviate poverty and social exclusion; on the contrary, it exacerbates them. The Washington consensus has been blind to this effect because corporate economists have traditionally excluded the social costs of economic activity from their models.[39] Similarly, most conventional economists have ignored the new economy's environmental cost—the increase and acceleration of global environmental destruction, which is as severe, if not more so, than its social impact.

The central enterprise of current economic theory and practice—the striving for continuing, undifferentiated economic growth—is clearly unsustainable, since unlimited expansion on a finite planet can only lead to catastrophe. Indeed, at the turn of this century it has become abundantly clear that our economic activities are harming the biosphere and human life in ways that may soon become irreversible.[40] In this precarious situation, it is paramount for humanity to systematically reduce its impact on the natural environment. As then-senator Al Gore declared courageously in 1992, "We must make the rescue of the environment the central organizing principle for civilization."[41]

Unfortunately, instead of following this admonition, the new economy has significantly increased our harmful impact on the biosphere. In *The Case Against the Global Economy*, Edward Goldsmith, founding editor of the leading European environmental journal *The Ecologist*, gives a succinct summary of the environmental impact of economic globalization.[42] He points out that the increase of environmental destruction with increasing economic growth is well illustrated by the examples of South Korea and Taiwan. During the 1990s, both countries achieved stunning rates of growth and were held up as economic models for the Third World by the World Bank. At the same time, the resulting environmental damage has been devastating.

In Taiwan, agricultural and industrial poisons have severely polluted nearly every major river. In some places, the water is not only devoid of fish and unfit to drink, but is actually combustible. The level of air pollution is twice that considered harmful in the United States; cancer rates have doubled since 1965, and the country has the world's highest incidence of hepatitis. In principle, Taiwan could use its new wealth to clean up its environment, but competitiveness in the global economy is so extreme that environmental regulations are eliminated rather than strengthened in order to lower the costs of industrial production.

One of the tenets of neoliberalism is that poor countries should concentrate on producing a few special goods for export in order to obtain foreign exchange, and should import most other commodities. This emphasis has led to the rapid depletion of the natural resources required to produce export crops in country after country—diversion of fresh water from vital rice paddies to prawn farms; a focus on water-intensive crops, such as sugar cane, that result in dried-up riverbeds; conversion of good agricultural land into cash-crop plantations; and forced migration of large numbers of farmers from their lands. All over the world there are countless examples of how economic globalization is worsening environmental destruction.[43]

The dismantling of local production in favor of exports and imports, which is the main thrust of the WTO's free-trade rules, dramatically increases the distance "from the farm to the table." In the United States, the average ounce of food now travels over a thousand miles before being eaten, which puts enormous stress on the environment. New highways and airports cut through primary forests; new harbors destroy wetlands and coastal habitats; and the increased volume of transport further pollutes the air and causes frequent oil and chemical spills. Studies in Germany have shown that the contribution of nonlocal food production to global warming is between six and twelve times higher than that of local production, due to increased CO_2 emissions.[44]

As ecologist and agricultural activist Vandana Shiva points out, the impact of climate instability and ozone depletion is born disproportionately by the South, where most regions depend on agriculture and where slight changes in climate can totally destroy rural livelihoods. In addition, many transnational corporations use the free-trade rules to relocate their resource-intensive and polluting industries in the South, thus further worsening environmental destruction. The net effect, in Shiva's words, is that "resources move from the poor to the rich, and pollution moves from the rich to the poor."[45]

The destruction of the natural environment in Third World countries goes hand in hand with the dismantling of rural people's traditional, largely self-sufficient ways of life, as American television programs and transnational advertising agencies promote glittering images of modernity to billions of people all over the globe without mentioning that the lifestyle of endless material consumption is utterly unsustainable. Edward Goldsmith estimates that, if all Third World countries were to reach the consumption level of the United States by the year 2060, the annual environmental damage from the resulting economic activities would be 220 times what it is today, which is not even remotely conceivable.[46]

Since money-making is the dominant value of global capitalism, its representatives seek to eliminate environmental regulations under the guise of free trade wherever they can, lest these regulations interfere with profits. Thus the new economy causes environmental destruction not only by increasing the impact of its operations on the world's ecosystems, but also by eliminating national environmental laws in country after country. In other words, environmental destruction is not only a side effect, but is also an integral part, of the design of global capitalism. "Clearly," Goldsmith concludes, "there is no way of protecting our environment within the context of a global 'free trade' economy committed to continued economic growth and hence to increasing the harmful impact of our activities on an already fragile environment."[47]

The Transformation of Power

The Information Technology Revolution has not only given rise to a new economy, but has also decisively transformed traditional relationships of power. In the Information Age, networking has emerged as a critical form of organization in all sections of society. Dominant social functions are increasingly organized around networks, and participation in these networks is a critical source of power. In this "network society," as Castells calls it, the generation of new knowledge, economic productivity, political and military power, and communication through the media are all connected to global networks of information and wealth.[48]

The rise of the network society has gone hand in hand with the decline of the nation-state as a sovereign entity.[49] Embedded in global networks of turbulent financial flows, governments are less and less able to control their national economic policies; they can no longer deliver the promises of the traditional welfare state; they are fighting a losing battle against a newly globalized criminal economy; and their authority and legitimization are increasingly called into question. In addition, the state is disintegrating from within through the corruption of the democratic process, as the political actors—especially in the United States—depend more and more on corporations and other lobbying groups, which finance the politicians' electoral campaigns in exchange for policies that favor their "special interests."

The emergence of a vast global criminal economy and its growing interdependence with the formal economy and with political institutions at all levels is one of the most disturbing features of the new network society. In their desperate attempts to escape marginality, individuals and groups who have been socially excluded become easy recruits for criminal organizations, which have established themselves in many poor neighborhoods and have become a significant social and cultural force in most parts of the world.[50] Crime, of course, is nothing new. But the global networking of powerful criminal organizations is a novel phenomenon that profoundly affects economic and political activities around the world, as Castells has documented in great detail.[51]

While drug traffic is the most significant operation of the global criminal networks, arms deals also play a significant role, in addition to the smuggling of goods and people, gambling, kidnapping, prostitution, counterfeiting of money and documents, and scores of other activities. The legalization of drugs would probably be the greatest threat to organized crime. However, as Castells notes wryly, "They can rely on the political blindness and misplaced morality of societies that do not come to terms with the bottom line of the problem: demand drives supply."[52]

Ruthless violence, often carried out by contracted killers, is an integral part of the criminal culture. As important, however, are the law-enforcement agents, judges, and politicians who are on the criminal organizations' payroll and who are sometimes cynically referred to as the "security apparatus" of organized crime.

Money laundering to the tune of hundreds of billions of dollars is the core activity of the criminal economy. The laundered money enters the formal economy through complex financial schemes and trade networks, thus introducing a destabilizing but unseen element into an already volatile system and making it even more difficult to control national economic policies. Financial crises may have been triggered by criminal activities in several parts of the world. In Latin America, by contrast, *narcotrafico* represents a secure and dynamic segment of regional and national economies. The Latin American drug industry is demand-driven, export-oriented, and fully internationalized. Unlike most of the legal trade, it is completely under Latin American control.

Like the business organizations in the formal economy, today's criminal organizations have restructured themselves as networks, both internally and in relation to each other. Strategic alliances have been formed between criminal organizations around the world, from the Colombian drug cartels to the Sicilian Mafia, the American Mafia, and the Russian criminal networks. New communication technologies, particularly mobile phones and laptop computers, are used widely to communicate and keep track of transactions. Thus Russian Mafia millionaires are now able to conduct their Moscow businesses online from safe California mansions while keeping a close eye on day-to-day operations.

According to Castells, the organizational strength of global crime is based on the "combination of flexible networking between local turfs, rooted in tradition and identity, in a favourable institutional environment, and the global reach provided by strategic alliances,"[53] Castells believes that today's criminal networks are probably more advanced than transnational corporations in their ability to combine local cultural identity and global business.

If the nation-state is losing its authority and legitimacy because of the pressures of the global economy and the undermining effects of global crime, what will take its place? Castells notes that political authority has been shifting to regional and local levels, and he speculates that this decentralization of power may give rise to a new kind of political organization, the "network state."[54] In a social network, different nodes may be of different sizes, and thus political inequalities and asymmetrical power relations will be common. However, all members of a network state are interdependent. When political decisions are made, their effects on any members, even the smallest, need to be taken into account, because they will necessarily affect the entire network.

The European Union may be the clearest manifestation of such a new network state. The regions and cities have access to it through their national governments, and they are also interconnected with one another horizontally through multiple partnerships across national boundaries. "The European Union does not supplant the existing nation-states," Castells concludes, "but, on the contrary, is a fundamental instrument for their survival on the condition of conceding shares of sovereignty in exchange for a greater say in the world."[55]

A similar situation exists in the corporate world. Today's corporations are increasingly organized as decentralized networks of smaller units; they are connected to networks of subcontractors, suppliers, and consultants; and units from different networks also form temporary strategic alliances and engage in joint ventures. In these network structures of ever varying geometries there are no real centers of power. By contrast, corporate power as a whole has increased enormously over the past few decades, as through never-ending mergers and acquisitions, the size of corporations continues to grow.

Over the past twenty years, transnational corporations have been extremely aggressive in extracting financial subsidies from the governments of the countries in which they operate, and in seeking to avoid paying taxes. They can be ruthless when it comes to ruining small businesses by undercutting their prices; they routinely withhold and distort information about potential dangers inherent in their products; and they have been very successful in coercing governments to eliminate regulatory constraints through free-trade agreements.[56]

Nevertheless, it would be false to think that a few megacorporations control the world. To begin with, real economic power has shifted to the global financial networks. Every corporation depends on what happens in those complex networks, which nobody controls. There are thousands of corporations today, all of whom compete and cooperate at the same time, and no individual corporation can dictate conditions.[57]

This diffusion of corporate power is a direct consequence of the properties of social networks. In a hierarchy, the exertion of power is a controlled, linear process. In a network it is a nonlinear process involving multiple feedback loops, and the results are often impossible to predict. The consequences of every action within the network spread throughout the entire structure, and any action that furthers a particular goal may have secondary consequences that conflict with that goal.

It is instructive to compare this situation with ecological networks. Although it may seem that in an ecosystem some species are more powerful than others, the concept of power is not appropriate, because nonhuman species (with the exception of some primates) do not force individuals to act in accordance with preconceived goals. There is dominance, but it is always acted out within a larger context of cooperation, even in predator-prey relationships.[58] The manifold species in an ecosystem do not form hierarchies, as is often erroneously stated, but exist in networks nested within networks.[59]

There is a crucial difference between the ecological networks of nature and the corporate networks in human society. In an ecosystem, no being is excluded from the network. Every species, even the smallest bacterium, contributes to the sustainability of the whole. In the human world of wealth and power, by contrast, large segments of the population are excluded from the global networks and are rendered economically irrelevant. The effects of corporate power on individuals and groups who are socially excluded are dramatically different from its effects on those who are members of the network society.

The Transformation of Culture

The communication networks that have shaped the new economy transmit not only information about financial transactions and investment opportunities, but also include global networks of news, the arts, science, entertainment, and other cultural expressions. These expressions, too, have been profoundly transformed by the Information Technology Revolution.[60]

Technology has made it possible to integrate communication by combining sounds and images with written and spoken words into a single "hypertext." Since culture is created and sustained by networks of human communications, it is bound to change with the transformation of its modes of communication.[61] Manuel Castells asserts that "the emergence of a new electronic communication system characterized by its global reach, its integration of all communication media, and its potential interactivity is changing and will change forever our culture."[62]

Like the rest of the corporate world, the mass media have increasingly evolved into global, decentralized network structures. This development was predicted in the 1960s by the visionary communications theorist Marshall McLuhan.[63] With his famous aphorism, "The medium is the message," McLuhan identified the unique nature of television and pointed out that, because of its seductiveness and powerful simulation of reality, it is the ideal medium for advertising and propaganda.

In most American households, radio and television have created a constant audiovisual environment that bombards the viewers and listeners with a never-ending stream of advertising messages. The entire programming of American network television is financed by and organized around its commercials, so that the communication of the corporate value of consumerism becomes television's overwhelming message. The coverage of the Olympic Games in Sydney by NBC was a crass example of an almost seamless mix of advertising and reporting. Instead of covering the Olympic Games, NBC chose to "produce" them for its viewers, packaging the programs in slick short segments, interspersed with commercials, in such a way that it was often difficult to distinguish between commercials and competitions. The images of athletes in competition were repeatedly transformed into schmaltzy symbols, and then reappeared in commercials just a few seconds later. As a result, the actual sports coverage was minimal.[64]

In spite of the constant barrage of advertising and the billions of dollars spent on it every year, studies have shown repeatedly that media advertising has virtually no specific impact on consumer behavior.[65] This startling discovery is further evidence for the observation that human beings, like all living systems, cannot be directed but can only be disturbed. As we have seen, choosing what to notice and how to respond is the very essence of being alive.[66]

This does not mean that the effects of advertising are negligible. Since the audiovisual media have become the principal channels for social and cultural communication in modern urban societies, people construct their symbolic images, values, and rules of behavior from the content offered by those media. Thus, companies and their products need to be present in the media to gain brand recognition. But how individuals will respond to a specific commercial is beyond the advertisers' control.

During the last two decades, new technologies have transformed the world of media to such an extent that many observers now believe that the era of mass media, in the traditional sense of limited contents sent to a homogeneous mass audience, will soon come to an end.[67] Major newspapers are now written, edited, and printed at a distance, with different editions tailored to regional markets appearing simultaneously. VCRs have become a major alternative to network television by making it possible to view videotaped movies and TV programs at convenient times. In addition, there has been an explosion of cable TV, satellite channels, and local community television stations.

The result of these technological innovations has been an extraordinary diversification of access to radio and television programs and, accordingly, a dramatic decline of network television audiences. In the United States, the three dominant TV networks captured 90 percent of the prime-time audience in 1980, but only 50 percent in 2000, and their share keeps shrinking. According to Castells, the current trend is clearly toward customized media for segmented audiences. Once people are able to receive a menu of media channels precisely tailored to their tastes, they will be willing to pay for it, which should eliminate advertising from these channels and may increase the quality of their programming.[68]

The rapid rise of pay-per-service television in the United States—HBO, Showtime, Fox Sports, etc.—does not mean that corporate control over television is diminishing. Although some of these channels are free of commercials, they are nevertheless controlled by corporations who will try to advertise in any way they can. The Internet, for example, has become the latest medium for massive corporate advertising. America Online (AOL), the leading Internet provider, is essentially a virtual shopping mall, saturated with ads. Although it offers Web access, its 20 million subscribers spend 84 percent of their time using AOL's in-house services and only 16 percent on the open Internet. And by merging with the media giant Time-Warner, AOL has added a huge arsenal of existing content and distribution channels to its domain, so that it can deliver its customers to major advertisers across a variety of media platforms.[69]

The media world today is dominated by a few giant multimedia conglomerates, like AOL-Time-Warner or ABC-Disney, which are vast networks of smaller companies with many kinds of interconnections and strategic alliances. Thus the media, like the corporate world as a whole, are becoming more decentralized and diversified, while the overall corporate impact on people's lives continues to increase.

The integration of all forms of cultural expression into a single electronic hypertext has not yet been realized, but the effects of such a development on our perceptions can already be gauged from the current contents of cable and network television programs and their associated web sites. The culture we create and sustain with our networks of communications includes not only our values, beliefs, and rules of conduct, but also our very perception of reality. As cognitive scientists have explained, human beings exist in language. By continually weaving a linguistic web, we coordinate our behavior and together bring forth our world.[70]

When this linguistic web becomes a hypertext of words, sounds, images, and other cultural expressions, mediated electronically and abstracted from history and geography, this is bound to influence profoundly the ways in which we see the world. As Castells points out, we can observe a pervasive blurring of levels of reality in the electronic media.[71] As different modes of communication borrow codes and symbols from each other, newscasts look more and more like talk shows, trial cases like soap operas, and reports on armed conflicts like action movies, and it becomes more and more difficult to distinguish the virtual from the real.

Since the electronic media, and especially television, have become the principal channels for communicating ideas and values to the public, politics is played out increasingly in the space of these media.[72] Media presence is as essential for politicians as it is for corporations and their products. In most societies, politicians who are not in the electronic networks of media communication do not stand a chance of gaining public support: they will remain simply unknown to the majority of voters.

With the blurring of news and entertainment, of information and advertising, politics becomes more and more like theater. The most successful politicians are no longer the ones with popular platforms, but those who come across well on television and who are adept at manipulating symbols and cultural codes. "Branding" candidates—i.e. making their names and images appealing by associating them firmly with seductive symbols in the viewers' minds—has become as important in politics as it is in corporate marketing. At a fundamental level, political power lies in the ability to use symbols and cultural codes effectively to frame political discourse in the media. As Castells emphasizes, this means that the power battles of the Information Age are cultural battles.[73]

The Question of Sustainability

In the last few years, the new economy's social and ecological impacts have been discussed extensively by scholars and community leaders, as has been documented in the preceding pages. Their analyses make it abundantly clear that global capitalism in its present form is unsustainable and needs to be fundamentally redesigned. Such a redesign is now advocated even by some "enlightened capitalists" who are worried about the highly volatile nature and self-destructive potential of the current system. Financier George Soros, who has been one of the most successful gamblers in the global casino, has recently begun to refer to the neoliberal doctrine of economic globalization as "market fundamentalism" and believes that it is as dangerous as any other kind of fundamentalism.[74]

In addition to its economic instability, the current form of global capitalism is ecologically and socially unsustainable, and hence not viable in the long run. Resentment against economic globalization is growing rapidly in all parts of the world. The ultimate fate of global capitalism may well be, as Manuel Castells puts it, "the social, cultural, and political rejection by large numbers of people around the world of an Automaton whose logic either ignores or devalues their humanity."[75] As we shall see, this rejection may already have begun.[76]

| six |

BIOTECHNOLOGY AT A TURNING POINT

When we think about advanced, twenty-first-century technologies, we tend to think not only about information technology but also about biotechnology. Like the Information Technology Revolution, the "biotech revolution" began with several decisive innovations in the 1970s and reached its initial climax in the 1990s.

Genetic engineering is sometimes considered as a special kind of information technology, since it involves the manipulation of genetic "information," but there are fundamental and very interesting differences between the conceptual frameworks underlying these two technologies. Whereas the understanding and use of networks has been at the very center of the Information Technology Revolution, genetic engineering is based on a linear and mechanistic building-block approach and has until very recently disregarded the cellular networks that are crucial to all biological functions.[1] As we move into the twenty-first century, it is fascinating to observe that the most recent advances in genetics are forcing molecular biologists to question many of the fundamental concepts on which their whole enterprise was originally based. This observation is the central theme of a brilliant evaluation of genetics at this turn of the century by biologist and science historian Evelyn Fox Keller whose arguments I shall follow through much of this chapter.[2]

Development of Genetic Engineering

Genetic engineering, in the words of molecular biologist Mae-Wan Ho, is "a set of techniques for isolating, modifying, multiplying, and recombining genes from different organisms."[3] It enables scientists to transfer genes between species that would never interbreed in nature, taking, for example, genes from a fish and putting them into a strawberry or a tomato, or putting human genes into cows or sheep, and thereby creating new "transgenic" organisms.

The science of genetics culminated in the discovery of the physical structure of DNA and the "breaking of the genetic code" during the 1950s,[4] but it took biologists another twenty years to develop two crucial techniques that made genetic engineering possible. The first, known as "DNA sequencing," is the ability to determine the exact sequence of genetic elements (the nucleotide bases) along any stretch of the DNA double helix. The second, "gene-splicing," is the cutting and joining together of pieces of DNA with the help of special enzymes isolated from microorganisms.[5]

It is important to understand that geneticists cannot insert foreign genes directly into a cell because of natural interspecies barriers and other protective mechanisms that break down or inactivate foreign DNA. To circumvent these obstacles, scientists splice the foreign genes first into viruses, or into viruslike elements that are routinely used by bacteria to trade genes.[6] These "gene transfer vectors" are then used to smuggle foreign genes into the selected recipient cells where the vectors, together with the genes spliced into them, insert themselves into the cell's DNA. If all the steps in this highly complex sequence work as planned, which is extremely rare, the result is a new transgenic organism. Another important gene-splicing technique is to produce copies of DNA sequences by inserting them into bacteria (again via transfer vectors), where they replicate rapidly.

The use of vectors to insert genes from the donor organism into the recipient organism is one of the main reasons why the process of genetic engineering is inherently hazardous. Aggressive infectious vectors could easily recombine with existing disease-causing viruses to generate new virulent strains. In her eye-opening book, *Genetic Engineering—Dream or Nightmare?*, Mae-Wan Ho speculates that the emergence of a host of new viruses and antibiotic resistances over the past decade may well be connected with the large-scale commercialization of genetic engineering during the same period.[7]

From the early days of genetic engineering, scientists have been aware of the dangers of inadvertently creating virulent strains of viruses or bacteria. In the 1970s and 1980s they took great care that the experimental transgenic organisms they created were contained in the laboratory, because they thought it unsafe to release them into the environment. In 1975 a group of concerned geneticists who gathered at Asilomar, California, issued the Asilomar Declaration, which called for a moratorium on genetic engineering until appropriate regulatory guidelines had been put in place.[8]

Unfortunately, this cautious and responsible attitude was largely abandoned during the 1990s, in a frantic rush to commercialize the newly developed genetic technologies in order to apply them in medicine and agriculture. At first, small biotech companies were organized around Nobel Prize winners at major American universities and medical research centers, and a few years later, these were bought by huge pharmaceutical and chemical corporations, which soon became aggressive proponents of biotechnology.

The 1990s saw several sensational announcements of genetic "cloning" of animals, including that of a sheep at the Roslin Institute in Edinburgh and of several mice at the University of Hawaii.[9] Meanwhile, plant biotechnology invaded agriculture with incredible speed. In the two years between 1996 and 1998 alone, the global area covered by transgenic crops increased more than tenfold, from 7 million to 74 million acres.[10] This massive release of genetically modified organisms (GMOs) into the environment added a new category of ecological risks to biotechnology's already existing problems.[11] Unfortunately, these risks are often waved aside by geneticists, who often have very little ecological knowledge or training.

As Mae-Wan Ho points out, genetic engineering techniques are now ten times faster and more powerful than they were twenty years ago; and new breeds of GMOs, designed to be ecologically vigorous, are deliberately released on a large scale, but in spite of greatly increased potential dangers, there have been no further joint declarations from geneticists calling for a moratorium. On the contrary, regulatory bodies have repeatedly given in to corporate pressures and have relaxed already inadequate safety regulations.[12]

As global capitalism began to thrive in the 1990s, its mentality of allowing money-making to supersede all other values engulfed biotechnology and seemed to sweep aside all ethical considerations. Many leading geneticists now either own biotech companies or have close ties to them. The overriding motivation for genetic engineering is not the advancement of science, the curing of disease, or the feeding of the hungry. It is the desire to secure unprecedented financial gain.

The biggest and perhaps most competitive enterprise in biotechnology so far has been the Human Genome Project—the attempt to identify and map the complete genetic sequence of the human species, which contains tens of thousands of genes. During the 1990s this effort turned into a fierce race between a government-funded project that made its discoveries available to the public and a private group of geneticists that kept its data secret in order to patent it and sell it to biotechnology companies. In its final dramatic phase, the race was decided by an unlikely hero, a young graduate student who singlehandedly wrote the decisive computer program that helped the public project win the race by three days, and thus prevented private control of the scientific understanding of human genes.[13]

The Human Genome Project began in 1990 as a collaborative program among several teams of top geneticists coordinated by James Watson (who, with Francis Crick, discovered the DNA double helix) and funded by the U.S. government to the tune of three billion dollars. A rough draft of the mapping was expected to be completed ahead of schedule in 2001, but while these efforts were under way Celera

Genomics, with superior computer power and funding from venture capitalists, overtook the government-sponsored project and patented its data to ensure exclusive commercial rights to the manipulation of human genes. In response, the public project (which had grown into an international consortium headed by geneticist Francis Collins) published its discoveries on the Internet on a daily basis to make sure that they were in the public domain and could not be patented.

By December 1999, the public consortium had identified 400,000 fragments of DNA, most of them smaller than an average gene, but they had no idea how to orient and assemble these pieces—"hardly worthy of being called a sequence," as their competitor, biologist Craig Venter, the founder of Celera Genomics, liked to observe. At this stage, David Haussler, a professor of computer science at the University of California at Santa Cruz, joined the consortium. Haussler believed that there was enough information in the collected data to design a special computer program that would assemble the pieces properly.

However, progress was painfully slow, and in May 2000, Haussler told one of his graduate students, James Kent, that the prospect of finishing ahead of Celera looked "grim." Like many scientists, Kent was very concerned that future work on understanding the human genome would be under the control of private corporations if the sequencing data could not be made public before it was patented. When he heard about the slow progress of the public project, he told his professor that he felt he could write an assembly program using a simpler and superior strategy.

Four weeks later, after working day and night and icing his wrists between long sessions of furious typing, James Kent had written 10,000 lines of code, completing the first assembly of the human genome. "He's unbelievable," Haussler told the *New York Times*. "This programme represents an amount of work that would have taken a team of five or ten programmers at least six months or a year. Jim [alone] in four weeks created . . . this extraordinarily complex piece of code."[14]

In addition to his assembly program, nicknamed the "golden path," Kent created another program, known as a browser, which enabled scientists to view the assembled sequence of the human genome for the first time and for free, without subscribing to Celera's database. The human genome race officially ended seven months later, when the public consortium and the Celera scientists published their results during the same week, the former in *Nature* and the latter in *Science*.[15]

Conceptual Revolution in Genetics

While the competition to map the human genome first raged, the very successes of these and of other DNA sequencing efforts triggered a conceptual revolution in genetics that is likely to show the futility of any hope that mapping the human genome will soon lead to tangible practical applications. In order to use genetic knowledge to influence the functioning of the organism—for example, to prevent or cure diseases—we need to know not only where specific genes are located, but also how they function. After sequencing major portions of the human genome and mapping the complete genomes of several plant and animal species, geneticists naturally turned their attention from gene structure to gene function; and when they did so, they realized how limited our knowledge of gene function still is. As Evelyn Fox Keller observes, "Recent developments in molecular biology have given us new appreciation of the magnitude of the gap between genetic information and biological meaning."[16]

For several decades after the discoveries of the DNA double helix and the genetic code, molecular biologists believed that the "secret of life" lay in the sequences of genetic elements along the DNA strands. If we could only identify and decode those sequences, the thinking went, we would understand the genetic "programs" that determine all biological structures and processes. Today, very few biologists still hold this belief. The newly developed sophisticated techniques of DNA sequencing and of related genetic research increasingly show that the traditional concepts of "genetic determinism"—including that of a genetic program, and maybe even the concept of the gene itself—are being seriously challenged and are in need of radical revision.

A profound shift of emphasis, from the structure of genetic sequences to the organization of metabolic networks, from genetics to epigenetics is taking place. It is a move from reductionist to systemic thinking. In the words of James Bailey, a geneticist at the Institute for Biotechnology in Zurich, "The current cascade of complete genome sequences . . . now compels a major shift in bioscience research toward integration and system behaviour."[17]

Stability and Change

To appreciate the magnitude and extent of this conceptual shift, we need to revisit the origins of genetics in Darwin's theory of evolution and Mendel's theory of heredity. When Charles Darwin formulated his theory in terms of the twin concepts of "chance variation" (later to be called random mutation) and natural selection, it soon became apparent that chance variations, as conceived by Darwin, could not explain the emergence of new characteristics in the evolution of species. Darwin shared with his contemporaries the assumption that the biological characteristics of an individual represented a blend of those of its parents, with both parents contributing more or less equal parts to the mixture. This meant that an offspring of a parent with a useful chance variation would inherit only 50 percent of the new characteristic, and would be able to pass on only 25 percent of it to the next generation. Thus the new characteristic would be diluted rapidly, with very little chance of establishing itself through natural selection.

Although the Darwinian theory of evolution introduced the radically new understanding of the origin and transformation of species that became one of the towering achievements of modern science, it could not explain the persistence of newly evolved traits, nor indeed the more general observation that each generation of living organisms, as it grows and develops, unfailingly displays the typical characteristics of its species. This remarkable stability applies even to particular individual features, such as clearly recognizable family resemblances that are frequently passed on faithfully from generation to generation.

Darwin himself recognized that the inability of his theory to explain the constancy of hereditary traits was a serious flaw for which he had no remedy. Ironically, the solution to his problem was discovered by Gregor Mendel only a few years after the publication of Darwin's *Origin of Species*, but was ignored for several decades until its rediscovery at the beginning of the twentieth century.

From his careful experiments with garden peas, Mendel deduced that there were "units of heredity"—later to be called genes—that did not blend in the process of reproduction, but were transmitted from generation to generation without changing their identity. With this discovery it could be assumed that random mutations would not disappear within a few generations but would be preserved, to be either reinforced or eliminated by natural selection.

With the discovery of the physical structure of genes by Watson and Crick in the 1950s, genetic stability became understood in terms of the faithful self-replication of the DNA double helix, and mutations, correspondingly, as occasional but very rare random errors in that process. Over subsequent decades, this understanding firmly established the concept of genes as clearly distinct and stable hereditary units.[18]

However, recent advances in molecular biology have now seriously challenged our understanding of genetic stability, and with it the entire image of genes as causal agents of biological life, which is deeply embedded in both popular and scientific thought. As Evelyn Fox Keller explains,

> To be sure, genetic stability remains as remarkable a property as ever, and it is clearly a property of all known organisms. The difficulty arises with the question of how that stability is maintained, and this has proven to be a far more complex matter than we could ever have imagined.[19]

When the chromosomes of a cell double themselves in the process of cell division, their DNA molecules divide in such a way that the two chains of the double helix separate, and each of them serves as a template for the construction of a new complementary chain. This self-replication takes place with amazing fidelity. The frequency of copying mistakes, or mutations, is roughly one in ten billion!

This extreme fidelity, which lies at the origin of genetic stability, is not just a consequence of the physical structure of DNA. In fact, a DNA molecule by itself is not able to self-replicate at all. It needs specific enzymes to facilitate every step of the self-replication process.[20] One kind of enzyme helps the two parent strands to unwind; another prevents the unwound strands from winding back together; and a host of further enzymes select the correct genetic elements, or "bases," for complementary binding, check the most recently added bases for accuracy, correct mismatches, and repair accidental damages to the DNA structure. Without this elaborate system of monitoring, proofreading, and repair, errors in the self-replication process would increase dramatically. Instead of one in ten billion, one in a hundred bases would be copied erroneously, according to current estimates.[21]

These recent discoveries show clearly that genetic stability is not inherent in the structure of DNA, but is an emergent property, resulting from the complex dynamics of the entire cellular network. In the words of Keller:

> The stability of gene structure thus appears not as a starting point but as an end-product—as the result of a highly orchestrated dynamic process requiring the participation of a large number of enzymes organized into complex metabolic networks that regulate and ensure both the stability of the DNA molecule and its fidelity in replication.[22]

When a cell replicates, it passes on not only the newly replicated DNA double helix, but also a full set of the necessary enzymes, as well as membranes and other cellular structures—in short, the entire cellular network. And thus the cellular metabolism continues without ever disrupting its self-generating network patterns.

In their attempts to understand the complex orchestration of the enzyme activity that gives rise to genetic stability, biologists recently were amazed to discover that the fidelity of DNA replication is not always maximized. There seem to be mechanisms that actively generate copying errors by relaxing some of the monitoring processes. Moreover, it appears that when and where mutation rates are increased in this way depends both on the organism and on the conditions in which the organism finds itself.[23] In every living organism, there is a subtle balance between genetic stability and "mutability"—the organism's ability actively to produce mutations.

The regulation of mutability is one of the most fascinating discoveries in current genetic research. According to Keller, this has become one of the hottest topics in molecular biology. "With the new analytical techniques that have now become available," she explains, "many aspects of the biochemical machinery involved in such regulation have been elucidated. But with every step toward elucidation, the picture is rendered ever more complex by the increasing wealth of detail."[24]

Whatever the specific dynamics of its regulation turn out to be, the implications of genetic mutability for our understanding of evolution are enormous. In the conventional neo-Darwinist view, DNA is seen as an inherently stable molecule subject to occasional random mutations, and evolution, accordingly, as being driven by pure chance, followed by natural selection.[25] The new discoveries in genetics will force biologists to adopt the radically different view that mutations are actively generated and regulated by the cell's epigenetic network, and that evolution is an integral part of the self-organization of living organisms. Molecular biologist James Shapiro wrote that:

> These molecular insights lead to new concepts of how genomes are organized and reorganized, opening a range of possibilities for thinking about evolution. Rather than being restricted to contemplating a slow process depending on random (i.e. blind) genetic variation . . . we are now free to think in realistic molecular ways about rapid genome restructuring guided by biological feedback networks.[26]

This new view of evolution as part of life's self-organization is further supported by extensive research in microbiology, which has shown that mutations are only one of three avenues of evolutionary change, the other two being the trading of genes between bacteria and the process of symbiogenesis—the creation of new forms of life through the merging of different species. The recent mapping of the human genome showed that many human genes originated from bacteria, confirming once more the theory of symbiogenesis proposed by microbiologist Lynn Margulis more than thirty years ago.[27] Taken together, these advances in genetics and microbiology amount to a dramatic conceptual shift in the theory of evolution—from the neo-Darwinist emphasis on "chance and necessity" to a systems view that sees evolutionary change as a manifestation of life's self-organization.

Since the systemic conception of life also identifies the self-organizing activity of living organisms with cognition,[28] this means that, ultimately, evolution must be seen as a cognitive process. As geneticist Barbara McClintock reflected prophetically in her 1983 Nobel lecture:

> In the future attention undoubtedly will be centred on the genome, and with greater appreciation of its significance as a highly sensitive organ of the cell, monitoring genomic activities and correcting common errors, sensing the unusual and unexpected events, and responding to them.[29]

Beyond Genetic Determinism

To summarize the first important insight from recent advances in genetic research: the stability of genes, the organism's "units of heredity," is not an intrinsic property of the DNA molecule but emerges from a complex dynamic of cellular processes. With this understanding of genetic stability, let us now turn to the central question of genetics: What do genes actually do? How do they give rise to characteristic hereditary traits and forms of behavior? After the discovery of the DNA double helix and the mechanism of its self-replication, it took molecular biologists another decade to find an answer to this question. Again, this research was spearheaded by James Watson and Francis Crick.[30]

To put it in greatly simplified terms, the cellular processes underlying biological forms and behavior are catalyzed by enzymes, and the enzymes are specified by genes. To produce a specific enzyme, the information encoded in the corresponding gene (i.e. the sequence of nucleotide bases along the DNA strand) is copied into a complementary RNA strand. The RNA molecule serves as a messenger, carrying the genetic information to a ribosome, the cellular structure where enzymes and other proteins are produced. At the ribosome, the genetic sequence is translated into instructions for assembling a sequence of amino acids, the basic building blocks of proteins. The celebrated genetic code is the precise correspondence by which successive triplets of genetic bases on the RNA strand are translated into a sequence of amino acids in the protein molecule.

With these discoveries the answer to the question of gene function seemed compellingly simple and elegant: genes encode the enzymes that are the necessary catalysts of all cellular processes. Thus genes determine biological traits and behavior, and each gene corresponds to a specific enzyme. This explanation has been called the Central Dogma of molecular biology by Francis Crick. It describes a linear causal chain from DNA to RNA, to proteins (enzymes) and to biological traits. In the colloquial paraphrase that has become popular among molecular biologists, "DNA makes RNA, RNA makes protein, and proteins make us."[31] The Central Dogma includes the assertion that its linear causal chain defines a one-way flow of information from the genes to the proteins, without the possibility of any feedback in the opposite direction.

The linear chain described by the Central Dogma is, in fact, far too simplistic to describe the actual processes involved in the synthesis of proteins. And the discrepancy between the theoretical framework and the biological reality is even greater when the linear sequence is shortened to its two end points, DNA and traits, so that the Central Dogma is turned into the statement, "Genes determine behavior." This view, known as genetic determinism, has become the conceptual basis of genetic engineering. It is promoted vigorously by the biotechnology industry and repeated constantly in the popular media: once we know the exact sequence of genetic bases in the DNA, we will understand how genes cause cancer, human intelligence, or violent behavior.

Genetic determinism has been the dominant paradigm in molecular biology for the past four decades, during which it has generated a host of powerful metaphors. DNA is often referred to as the organism's genetic "program" or "blueprint," or as the "book of life," and the genetic code as the universal "language of life." As Mae-Wan Ho points out, the exclusive focus on genes has almost completely eclipsed the organism from the biologists' view. The living organism tends to be regarded simply as a collection of genes, while it is totally passive, subject to random mutations and selective forces in the environment over which it has no control.[32]

According to molecular biologist Richard Strohman, the basic fallacy of genetic determinism lies in a confusion of levels. A theory that worked well, at least initially, for understanding the genetic code—how genes encode information for the production of proteins—has been extended to a theory of life that views genes as causal agents of all biological phenomena. "We are mixing our levels in biology and it doesn't work," he concludes. "The illegitimate extension of a genetic paradigm from a relatively simple level of genetic coding and decoding to a complex level of cellular behaviour represents an epistemological error of the first order."[33]

Problems with the Central Dogma

The problems with the Central Dogma became apparent during the late 1970s, when biologists extended their genetic research beyond bacteria. They soon found out that in higher organisms the simple correspondence between DNA sequences and sequences of amino acids in proteins no longer exists, and that the elegant principle of "one gene—one protein" has to be abandoned. Indeed, it seems—perhaps not unreasonably—that the processes of protein synthesis become increasingly complex as we move to more complex organisms.

In higher organisms, the genes that code for proteins tend to be fragmented rather than form continuous sequences.[34] They consist of coding segments interspersed with long repetitive noncoding sequences whose function is still unclear. The proportion of coding DNA varies a great deal and in some organisms can be as low as 1 to 2 percent. The rest is often referred to as "junk DNA." However, since natural selection has preserved these noncoding segments throughout the history of evolution, it is reasonable to assume that they play an important though still mysterious role.

Indeed, the complex genetic landscape revealed by the mapping of the human genome contains some intriguing clues about human evolution—a kind of genetic fossil record consisting of "jumping genes" that broke away from their chromosomes in our distant evolutionary past, replicated themselves independently, and then reinserted their copies into various sections of the main genome. Their distribution indicates that some of these noncoding sequences may contribute to the overall regulation of genetic activity.[35] In other words, they are not junk at all.

When a fragmented gene is copied into an RNA strand, the copy must be processed before the assembly of the protein can begin. Special enzymes come into play that remove the noncoding segments and then splice the remaining coding segments together to form a mature transcript: the messenger RNA is edited on its way to protein synthesis.

This editing process is not unique: the coding sequences can be spliced together in more than one way, and each alternative splicing will result in a different protein. Thus, many different proteins can be produced from the same primary genetic sequence, sometimes as many as several hundred according to current estimates.[36] This means that we have to give up the principle that each gene leads to the production of a specific enzyme (or other protein). Which enzyme is produced can no longer be deduced from the genetic sequence in the DNA. Keller states that:

> The signal (or signals) determining the specific pattern in which the final transcript is to be formed . . . [comes from] the complex regulatory dynamics of the cell as a whole . . . Unravelling the structure of such signalling pathways has become a major focus of contemporary molecular biology.[37]

Another recent surprise has been the discovery that the regulatory dynamics of the cellular network determine not only which protein will be produced from a given fragmented gene, but also how this protein will function. It has been known for some time that a protein can function in many different ways, depending on its context. Now scientists have discovered that the complex three-dimensional structure of a protein molecule can be changed by a variety of cellular mechanisms, and that these changes alter the molecule's function.[38] In short, cellular dynamics may lead to the emergence of many proteins from a single gene and of many functions from a single protein—a far cry indeed from the linear causal chain of the Central Dogma.

When we shift our attention from a single gene to the entire genome, and correspondingly from the making of a protein to the making of the whole organism, we encounter a different set of problems with genetic determinism. For example, when cells divide in the development of an embryo, each new cell receives exactly the same set of genes, and yet the cells specialize in very different ways, becoming muscle cells, blood cells, nerve cells, and so on. Developmental biologists concluded from this observation many decades ago that cell types differ from one another not because they contain different genes, but because different genes are active in them. In other words, the structure of the genome is the same in all these cells, but the patterns of gene activity are different. The question, then, is: What causes the differences in gene activity, or gene "expression," as it is technically known? As Keller puts it, "Genes do not simply *act:* they must be *activated.*"[39] They are turned on and off in response to specific signals.

A similar situation arises when we compare the genomes of different species. Recent genetic research has revealed surprising similarities between the genomes of humans and chimpanzees, and even between those of humans and mice. Geneticists now believe that the basic body plan of an animal is built from very similar sets of genes across the entire animal kingdom.[40] And yet the result is a great variety of radically different creatures. The differences, again, seem to lie in the patterns of gene expression.

To solve the problem of gene expression, molecular biologists

François Jacob and Jacques Monod in the early 1960s very ingeniously introduced a distinction between "structural genes" and "regulator genes." The structural genes, they maintained, are the ones that code for proteins, while the regulator genes control the rates of DNA transcription and thereby regulate gene expression.[41]

By assuming that these regulatory mechanisms are themselves genetic, Jacob and Monod managed to stay within the paradigm of genetic determinism, and they emphasized this point by using the metaphor of a "genetic program" to describe the process of biological development. Since computer science was establishing itself as an exciting, avant-garde discipline at the same time, the metaphor of the genetic program was very powerful and quickly became the dominant way of explaining biological development.

Subsequent research has shown, however, that the program for activating genes does not reside in the genome, but in the cell's epigenetic network. A number of cellular structures that are involved in regulating gene expression have been identified. They include structural proteins, hormones, networks of enzymes and many other molecular complexes. In particular, the "chromatin"—a large number of proteins that are tightly intertwined with the DNA strands inside the chromosomes—seems to play a critical role, as it constitutes the genome's most immediate environment.[42]

What emerges is the growing realization that the biological processes involving genes—the fidelity of DNA replication, the rate of mutations, the transcription of coding sequences, the selection of protein functions, and the patterns of gene expression—are all regulated by the cellular network in which the genome is embedded. This network is highly nonlinear, containing multiple feedback loops, so that patterns of genetic activity continually change in response to changing circumstances.[43]

DNA is an essential part of the epigenetic network, but it is not the sole causal agent of biological forms and functions as the Central Dogma would have it. Biological form and behavior are emergent properties of the network's nonlinear dynamics, and we can expect that our understanding of these processes of emergence will increase significantly when complexity theory is applied to the new discipline of "epigenetics." Indeed, this approach is currently being pursued by several biologists and mathematicians.[44]

Complexity theory may also shed new light on an intriguing property of biological development that was discovered almost a hundred years ago by the German embryologist Hans Driesch. With a series of careful experiments on sea urchin eggs, Driesch showed that he could destroy several cells in the very early stages of the embryo, and it would still grow into a full, mature sea urchin.[45] Similarly, more recent genetic experiments have shown that knocking out single genes, even when they were thought to be essential, had very little effect on the functioning of the organism.[46]

The very remarkable stability and robustness of biological development means that an embryo may start from different initial stages—for example, if single genes or entire cells are destroyed accidentally—but will nevertheless reach the same mature form that is characteristic of its species. Evidently, this phenomenon is quite incompatible with genetic determinism. The question is, in Keller's words, "What keeps development on track?"[47]

There is an emerging consensus among genetic researchers that this robustness indicates a functional redundancy in genetic and metabolic pathways. It seems that cells maintain multiple pathways for the production of essential cellular structures and the support of essential metabolic processes.[48] This redundancy ensures not only the remarkable stability of biological development but also great flexibility and adaptability to unexpected environmental changes. Genetic and metabolic redundancy may be seen, perhaps, as the equivalent of biodiversity in ecosystems. It seems that life has evolved ample diversity and redundancy at all levels of complexity.

The observation of genetic redundancy is in stark contradiction to genetic determinism, and in particular to the metaphor of the "selfish gene" proposed by biologist Richard Dawkins.[49] According to Dawkins, genes behave as if they were selfish by constantly competing, via the organisms they produce, to leave more copies of themselves. From this reductionist perspective, the widespread existence of redundant genes makes no evolutionary sense. From the systemic point of view, by contrast, we recognize that natural selection operates not on individual genes but on the organism's patterns of self-organization. As Keller puts it, "It is the endurance of the life cycle itself that has . . . become the subject of evolution."[50]

The existence of multiple pathways is, of course, an essential property of all networks; it may even be seen as the defining characteristic of a network. It is therefore not surprising that nonlinear dynamics (the mathematics of complexity theory), which is eminently suited to the analysis of networks, should contribute important insights into the nature of developmental robustness and stability.

In the language of complexity theory, the process of biological development is seen as a continuous unfolding of a nonlinear system as the embryo forms out of an extended domain of cells.[51] This "cell sheet" has certain dynamical properties that give rise to a sequence of deformations and foldings as the embryo emerges. The entire process can be represented mathematically by a trajectory in "phase space" moving inside a "basin of attraction" toward an "attractor" that describes the functioning of the organism in its stable adult form.[52]

A characteristic property of complex nonlinear systems is that they display a certain "structural stability." A basin of attraction can be disturbed or deformed without changing the system's basic characteristics. In the case of a developing embryo this means that the initial conditions of the process can be changed to some extent without seriously disturbing development as a whole. Thus developmental stability, which seems quite mysterious from the perspective of genetic determinism, is recognized as a consequence of a very basic property of complex nonlinear systems.

What Is a Gene?

The amazing progress made by geneticists in their efforts to identify and sequence particular genes and to map entire genomes has brought with it an increasing awareness that we need to go beyond genes if we really want to understand genetic phenomena. It may well be that we will be forced to abandon the concept of the gene altogether. As we have seen, genes are certainly not the independent and distinct causal agents of biological phenomena postulated by genetic determinism, and even their structure seems to elude precise definition.

Geneticists even find it difficult to agree on how many genes the human genome contains, because the portion of genes that code for amino acid sequences seems to be less than 2 percent. And as these coding genes are fragmented, interspersed by long noncoding sequences, the answer to the question where a specific gene begins and ends is anything but easy. Before the completion of the Human Genome Project, estimates of the total number of genes ranged between 30,000 and 120,000. Now it looks as if the lower end of this range is closer to the actual number, but not all geneticists agree.

It may well turn out that all we can say about genes is that they are continuous or discontinuous DNA segments whose precise structures and specific functions are determined by the dynamics of the surrounding epigenetic network and may change with changing circumstances. Geneticist William Gelbart goes even further when he writes:

> Unlike chromosomes, genes are not physical objects but are merely concepts that have acquired a great deal of historic baggage over the past decades . . . We may well have come to the point where the use of the term "gene" is of limited value and might in fact be a hindrance to our understanding of the genome.[53]

In her extensive review of the current state of genetics, Evelyn Fox Keller comes to a similar conclusion:

> Even though the message has yet to reach the popular press, to an increasingly large number of workers at the forefront of contemporary research, it seems evident that the primacy of the gene as the core explanatory concept of biological structure and function is more a feature of the twentieth century than it will be of the twenty-first.[54]

The fact that many of the leading researchers in molecular genetics now realize the need to go beyond genes and adopt a wider epigenetic perspective is important when we try to assess the current state of biotechnology. We shall see that the problems with the understanding of the relationship between genes and disease, the use of cloning in medical research and the applications of biotechnology to agriculture are all rooted in the narrow conceptual framework of genetic determinism and are likely to persist until a broader systemic view has been embraced by biotechnology's main proponents.

Genes and Disease

When the techniques of DNA sequencing and gene splicing were developed in the 1970s, the new biotech companies and their geneticists first turned to the medical applications of genetic engineering. Since genes were thought to determine biological functions, it was natural to assume that the root causes of biological disorders could be found in genetic mutations, and so geneticists set themselves the task of precisely identifying the genes that caused specific diseases. If they were successful in doing so, they thought, they might be able to prevent or cure these "genetic" diseases by correcting or replacing the defective genes.

The biotechnology companies saw the development of such genetic therapies as a tremendous business opportunity, even if actual therapeutic successes would lie far in the future, and began to promote vigorously their genetic research in the media. Year after year, bold headlines in newspapers and cover stories in magazines excitedly reported discoveries of new "disease-causing" genes and corresponding new potential therapies, usually with serious scientific caveats appearing a few weeks later but published as small notices among the bulk of other news.

Geneticists soon discovered that there is a huge gap between the ability to identify genes that are involved in the development of disease and the understanding of their precise function, let alone their manipulation to obtain a desired outcome. As we now know, this gap is a direct consequence of the mismatch between the linear causal chains of genetic determinism and the nonlinear epigenetic networks of biological reality.

The evocative term "genetic engineering," means that the public usually assumes that the manipulation of genes is an exact, well-understood mechanical procedure. Indeed, it is usually presented as such in the popular press. In the words of biologist Craig Holdrege:

> We hear of genes being *cut* or *spliced* by enzymes, and of new DNA combinations being *manufactured* and *inserted* into the cell. The cell incorporates the DNA into its *machinery*, which begins to *read information* that is *encoded* in the new DNA. This *information* is then *expressed* in the *manufacture* of corresponding proteins that have a particular function in the organism. And so, as if resulting from such precisely determinate procedures, the transgenic organism takes on new traits.[55]

The reality of genetic engineering is much more messy. At the current state of the art, geneticists cannot control what happens in the organism. They can insert a gene into the nucleus of a cell with the help of a specific gene transfer vector, but they never know whether the cell will incorporate it into its DNA, nor where the new gene will be located, nor what effects this will have on the organism. Thus, genetic engineering proceeds by trial and error in a way that is extremely wasteful. The average success rate of genetic experiments is only about 1 percent, because the living background of the host organism, which determines the outcome of the experiment, remains largely inaccessible to the engineering mentality that underlies our current biotechnologies.[56]

"Genetic engineering," explains biologist David Ehrenfeld, "is based on the premise that we can take a gene from species A, where it does some desirable thing, and move it into species B, where it will continue to do that same desirable thing. Most genetic engineers know that this is not always true, but the biotech industry as a whole acts as if it were."[57] Ehrenfeld points out that this premise encounters three main problems.

First, gene expression depends on the genetic and cellular environment (the whole epigenetic network) and can change when genes are put into a new environment. "Time and time again," writes molecular biologist Richard Strohman, "we find that genes associated with diseases of mice have no such association with those genes in humans . . . It appears, therefore, that mutation even in key genes will or will not have an effect, depending on the genetic background in which it finds itself."[58]

Second, genes usually have multiple effects, and undesirable effects that are suppressed in one species may be expressed when the gene is transferred to another species. And third, many traits involve multiple genes, perhaps even on different chromosomes, which are very resistant to being manipulated. Taken together, these three problems are the reason why the medical applications of genetic engineering have so far not yielded the desired results. As David Weatherall, director of Oxford University's Institute of Molecular Medicine, sums up, "Transferring genes into a new environment and enticing them to . . . do their jobs, with all the sophisticated regulatory mechanisms that are involved, has, so far, proved too difficult a task for molecular geneticists."[59]

Initially, geneticists hoped to associate specific diseases with single genes, but it turned out that single-gene disorders are extremely rare, accounting for less than 2 percent of all human diseases. Even in these clear-cut cases—for example, sickle-cell anemia, muscular dystrophy, or cystic fibrosis—where a mutation causes a malfunction in a single protein of crucial importance, the links between the defective gene and the onset and course of the disease are still poorly understood. The development of sickle-cell anemia, for example, which is common in Africans and African-Americans, can be dramatically different in individuals carrying the same defective gene, varying from early childhood death to a virtually unrecognized condition in middle age.[60]

Another problem is that the defective genes in these single-gene diseases are often very, very large. The gene that is critical to cystic fibrosis, a disease common among Northern Europeans, consists of some 230,000 base pairs and codes for a protein composed of almost 1,500 amino acids. More than 400 different mutations have been observed in this gene. Only one of them results in the disease, and identical mutations may lead to different symptoms in different individuals. All this makes screening for the "cystic fibrosis defect" highly problematic.[61]

The problems encountered in the rare single-gene disorders are compounded when geneticists study common diseases like cancer and heart disease, which involve networks of multiple genes. In these cases, observes Evelyn Fox Keller:

> the limits of current understanding are far more conspicuous. The net effect is that, while we have become extraordinarily proficient at identifying genetic risks, the prospect of significant medical benefits—benefits that only a decade ago were expected to follow rapidly upon the heels of the new diagnostic techniques—recedes ever further into the future.[62]

This situation is unlikely to change until geneticists begin to go beyond genes and focus on the complex organization of the cell as a whole. As Richard Strohman explains:

> In the case of coronary artery disease, [for example], there are more than 100 genes identified as having some interactive contribution. With networks of 100 genes and their products interacting with subtle environments to affect [biological functions], it is naive to think that some kind of nonlinear networking theory could be omitted from a diagnostic analysis.[63]

In the meantime, however, biotechnology companies continue to promote the outdated dogma of genetic determinism to justify their research. As Mae-Wan Ho points out, their attempts to identify genetic predispositions for diseases like cancer, diabetes, or schizophrenia—and worse, for conditions such as alcoholism or criminality—stigmatizes individuals and diverts attention from the crucial role of social and environmental factors that affect these conditions.[64]

The primary interest of the biotech companies, of course, is not human health or progress in medicine, but financial gain. One of the most effective ways of ensuring that the shareholder values of their ventures remain high, despite the lack of any significant medical benefits, is to perpetuate the perception among the general public that genes determine behavior.

The Biology and Ethics of Cloning

Genetic determinism has also decisively shaped public discussions of cloning after the recent dramatic successes in growing new organisms by genetic manipulation rather than sexual reproduction. The procedure used in these cases is different from cloning in the strict sense of the term, as we shall see below, but is now commonly described as "cloning" in the press.[65]

When the news became public in 1997 that a sheep had been "cloned" in this way by embryologist Ian Wilmut and his colleagues at the Roslin Institute in Scotland, it not only generated instant acclaim from the scientific community, but also aroused intense anxieties and public debates. Was the cloning of human beings now imminent, people wondered? Were there any ethical guidelines? Why had this research been allowed to go on, sheltered from public review, in the first place?

As evolutionary biologist Richard Lewontin points out in a thoughtful review of the science and ethics of cloning, the whole controversy needs to be understood against the background of genetic determinism.[66] Since the general public is unaware of the basic fallacy of the doctrine that genes "make" the organism, it naturally tends to believe that identical genes make identical people. In other words, most people confuse the genetic state of an organism with the totality of the biological, psychological, and cultural characteristics of a human being. Much more than genes is involved in the development of an individual—both in the emergence of biological form and in the formation of a unique human personality from certain life experiences. Hence, the notion of "cloning Einstein" is absurd.

As we shall see below, identical twins are genetically much more identical than a cloned organism is to its gene donor, and yet their personalities and life histories are usually quite different, in spite of the efforts of many parents to enforce the similarities between their twins by dressing them identically, giving them the same education, and so on. Any fears that cloning would violate an individual's unique identity are unfounded. In the words of Lewontin, "The question . . . is not whether genetic identity per se destroys individuality, but whether the erroneous state of public understanding of biology will undermine an individual's own sense of uniqueness and autonomy."[67] However, I need to add right away that the cloning of human beings would be morally reprehensible and unacceptable for other reasons, which I shall address.

Genetic determinism also supports the view that there might be justifiable motivations for cloning human beings in certain special circumstances—a woman, for example, whose husband is in a fatal coma after an accident, and who desperately wants a child by him; or a sterile man whose entire family has been killed and who does not want to see his biological heritage become extinct. Underlying these hypothetical cases is always the flawed assumption that preserving a person's genetic identity means, somehow, preserving his or her very essence. Interestingly, as Lewontin points out, this belief is a continuation of the ancient association of human blood with characteristics of social class or individual personality. Throughout the centuries, this erroneous association has generated a host of spurious moral problems and given rise to countless tragedies.

The real ethical questions about cloning become apparent when we understand the genetic manipulations involved in the current practices and the motivations behind this research. When biologists attempt to "clone" an animal today, they take an adult egg from one animal, remove its nucleus, and fuse the remaining cell with a nucleus (or an entire cell) from another animal. The resulting "hybrid" cell, the equivalent of a fertilized egg, is then developed in vitro and, after making sure that it is developing "normally," is implanted in the womb of a third animal, which serves as a surrogate mother and carries the embryo to term.[68] The scientific achievement of Wilmut and his colleagues was to demonstrate that the obstacle of cell specialization can be overcome. Adult cells of an animal are specialized, and their reproduction normally will only result in more cells of the same kind. Biologists had assumed that this specialization was irreversible. The scientists at the Roslin Institute showed that, somehow, it can be reversed by the interactions between the genome and the cellular network.

Unlike identical twins, the "cloned" animal is not completely identical, genetically, to its gene donor, because the manipulated cell from which it grew was composed not only of the nucleus from one donor, which provided the bulk of the genome, but also of the enucleated cell from another donor, which contained additional genes outside its nucleus.[69]

The real ethical problems surrounding the current cloning procedure are rooted in the biological developmental problems it generates. They are a consequence of the crucial fact that the manipulated cell from which the embryo grows is a hybrid of cellular components from two different animals. Its nucleus stems from one organism, while the rest of the cell, which contains the entire epigenetic network, stems from another. Because of the enormous complexity of the epigenetic network and its interactions with the genome, the two components will only very rarely be compatible, and our knowledge of cellular regulatory functions and signalling processes is still far too limited to know how to make them compatible. Thus, the currently practiced cloning procedure is based much more on trial and error than on an understanding of the underlying biological processes. In the Roslin Institute experiment, 277 embryos were created, but only one "cloned" sheep survived—a success rate of about one third of 1 percent.

Besides the question of whether so many embryos should be wasted in the interest of science, we also need to consider the nature of the nonviable creatures that are generated. In natural reproduction, the cells in the developing embryo divide in such a way that the processes of cell division and chromosome (and DNA) replication are in perfect synchrony. This synchrony is part of the cellular regulation of genetic activity.

In the case of "cloning," by contrast, the chromosomes may easily divide out of synchrony with the division of the embryonic cells because of incompatibilities between the two components of the initial manipulated cell.[70] This will result in either additional or missing chromosomes, so that the embryo will be abnormal. It may either die, or worse, may develop some monstrous growth. To use animals in such a way would raise ethical questions even if the research were motivated entirely by the desire to increase medical knowledge and help humanity. In the current situation these questions are much more urgent because the pace and direction of research are determined mainly by commercial interests.

The biotechnology industry is pursuing numerous projects in which cloning techniques are used for potential financial gain even though the health risks are often high and the benefits questionable. One line of research is to produce animal embryos whose cells and tissues might be useful for human therapeutic purposes. Another is to insert mutated human genes into animals so that they can serve as models for human diseases. For example, mice have been engineered to develop cancer, and the resulting sick transgenic animals have been patented![71] It is not surprising that most people feel a sense of revulsion about these business ventures.

Another major biotechnology project is to modify genetically domestic livestock in such a way that their milk contains useful drugs. As in the research projects mentioned above, these efforts require that many embryos be manipulated and discarded before a few transgenic animals are produced, and even those are often very sick. In addition, the question of whether the end product is safe for human consumption is paramount in the case of transgenic milk. Since genetic engineering always involves infectious gene transfer vectors that can easily recombine to create new pathogenic viruses, the hazards of transgenic milk far outweigh any potential benefits.[72]

The ethical problems of cloning experiments on animals would be magnified enormously were they to involve human beings. How many human embryos would we be prepared to sacrifice? How many developmental monstrosities would we allow to be created in such Faustian research? It is evident that any attempt to clone human beings at this stage of our knowledge would be totally immoral and unacceptable. Indeed, even in the case of cloning experiments on animals it is the moral duty of the scientific community to establish strict ethical guidelines and open its research to full public review.

Biotechnology in Agriculture

The applications of genetic engineering to agriculture have aroused much more widespread resistance among the general public than have the medical applications. There are several reasons for this resistance, which has grown into a worldwide political movement within the past few years. Most people around the world have a very basic existential relationship to food and are naturally worried when they feel that their food has been chemically contaminated or genetically manipulated. Even though they may not understand the complexities of genetic engineering, they become suspicious when they hear about new food technologies being developed in secret by powerful corporations who try to sell their products without any health warnings, labels, or even discussions. In recent years the glaring gap between the advertisements of the biotech industry and the realities of food biotechnology has become all too apparent.

The biotech ads portray a brave new world in which nature will be brought under control. Its plants will be genetically engineered commodities, tailored to customers' needs. New crop varieties will be drought tolerant and resistant to insects and weeds. Fruits will not rot or bruise. Agriculture will no longer be dependent on chemicals and hence will no longer damage the environment. Food will be better and safer than ever before, and world hunger will disappear.

Environmentalists and social justice advocates feel a strong sense of déjà vu when reading or hearing such optimistic but utterly naïve projections of the future. Many of us remember vividly that very similar language was used by the same agrochemical corporations when they promoted a new era of chemical farming, hailed as the "Green Revolution," several decades ago.[73] Since that time, the dark side of chemical agriculture has become painfully evident.

It is well known today that the Green Revolution has helped neither farmers, nor the land, nor the consumers. The massive use of chemical fertilizers and pesticides changed the whole fabric of agriculture and farming, as the agrochemical industry persuaded farmers that they could make money by planting large fields with a single highly profitable crop and by controlling weeds and pests with chemicals. This practice of single-crop monoculture entailed a high risk of large acreages being destroyed by a single pest, and it also seriously affected the health of farm workers and people living in agricultural areas.

With the new chemicals, farming became mechanized and energy intensive, favoring large corporate farmers with sufficient capital, and forcing most of the traditional single-family farmers to abandon their land. All over the world, large numbers of people have left rural areas and joined the masses of urban unemployed as victims of the Green Revolution.

The long-term effects of excessive chemical farming have been disastrous for the health of the soil and for human health, for our social relations, and for the entire natural environment on which our well-being and future survival depends. As the same crops were planted and fertilized synthetically year after year, the balance of the ecological processes in the soil was disrupted; the amount of organic matter diminished, and with it the soil's ability to retain moisture. The resulting changes in soil texture entailed a multitude of interrelated harmful consequences—loss of humus, dry and sterile soil, wind and water erosion, and so on.

The ecological imbalance caused by monocultures and excessive use of chemicals also resulted in enormous increases in pests and crop diseases, which farmers countered by spraying ever larger doses of pesticides in vicious cycles of depletion and destruction. The hazards for human health increased accordingly as more and more toxic chemicals seeped through the soil, contaminated the water table and showed up in our food.

Unfortunately, it seems that the agrochemical industry has not learned the lessons of the Green Revolution. According to biologist David Ehrenfeld:

> Like high-input agriculture, genetic engineering is often justified as a humane technology, one that feeds more people with better food. Nothing could be further from the truth. With very few exceptions, the whole point of genetic engineering is to increase the sales of chemicals and bio-engineered products to dependent farmers.[74]

The simple truth is that most innovations in food biotechnology have been profit-driven rather than need-driven. For example, soybeans were engineered by Monsanto to be resistant specifically to the company's herbicide Roundup so as to increase the sales of that product. Monsanto also produced cotton seeds containing an insecticide gene in order to boost seed sales. Technologies like these increase farmers' dependence on products that are patented and protected by "intellectual property rights," which make the age-old farming practices of reproducing, storing, and sharing seeds illegal. Moreover, the biotech companies charge "technology fees" in addition to the seed price or force farmers to pay inflated prices for seed-herbicide packages.[75]

Through a series of massive mergers and because of the tight control afforded by genetic technologies, an unprecedented concentration of ownership and control over food production is now under way.[76] The top ten agrochemical companies control 85 percent of the global market; the top five control virtually the entire market for genetically modified (GM) seeds. Monsanto alone bought into the major seed companies in India and Brazil, in addition to buying numerous biotech companies, while Du Pont bought Pioneer Hi-Bred, the world's largest seed company. The goal of these corporate giants is to create a single world agricultural system in which they would be able to control all stages of food production and manipulate both food supplies and prices. As a Monsanto executive explained, "What you are seeing is a consolidation of the entire food chain."[77]

The leading agrochemical corporations all plan to introduce versions of the "terminator technology"—plants with genetically sterilized seeds that would force farmers to buy patented products year after year and end their vital ability to develop new crops. This would be especially devastating in the Southern Hemisphere, where 80 percent of crops are grown from saved seed. More than anything else, these plans expose the stark commercial motivations behind GM foods. Many scientists working for these corporations may sincerely believe that their research will help to feed the world and improve the quality of our food, but they operate within a culture of power and control with an inability to listen and with narrow reductionist views, in which ethical concerns are not part of corporate strategies.

Biotechnology proponents have argued repeatedly that GM seeds are crucial to feed the world, using the same flawed reasoning that was advanced for decades by the proponents of the Green Revolution. Conventional food production, they maintain, will not keep pace with the growing world population. Monsanto's ads proclaimed in 1998: "Worrying about starving future generations won't feed them. Food biotechnology will."[78] As agroecologists Miguel Altieri and Peter Rosset point out, this argument is based on two erroneous assumptions.[79] The first is that world hunger is caused by a global shortage of food; the second is that genetic engineering is the only way to increase food production.

Development agencies have known for a long time that there is no direct relationship between the prevalence of hunger and a country's population density or growth. There is widespread hunger in densely populated countries like Bangladesh and Haiti, but also in sparsely populated ones like Brazil and Indonesia. Even in the United States, in the midst of super-abundance, there are between 20 and 30 million malnourished people.

In their classic study, "World Hunger: Twelve Myths," now published in an updated edition, development specialists Frances Moore Lappé and her colleagues at the Institute for Food and Development

Policy gave a detailed account of world food production that surprised many readers.[80] They showed that abundance, not scarcity, best describes the food supply in today's world. During the past three decades, increases in global food production have outstripped world population growth by 16 percent. During that time, mountains of surplus grain have pushed prices strongly downward on world markets. Increases in food supplies have kept ahead of population growth in every region except Africa during the past fifty years. A 1997 study found that in the developing world, 78 percent of all malnourished children under five live in countries with food surpluses. Many of these countries, in which hunger is rampant, export more agricultural goods than they import.

These statistics clearly show that the argument that biotechnology is needed to feed the world is highly disingenuous. The root causes of hunger around the world are unrelated to food production. They are poverty, inequality, and lack of access to food and land.[81] People go hungry because the means to produce and distribute food are controlled by the rich and powerful: world hunger is not a technical but a political problem. When agrobusiness executives assert that it will persist unless their latest biotechnologies are adopted, Miguel Altieri points out that they ignore the social and political realities. "If the root causes are not addressed," he retorts, "hunger will persist no matter what technologies are used."[82]

Biotechnology, of course, could have a place in agriculture in the future if it were used judiciously in conjunction with appropriate social and political measures, and if it could help produce better food without any harmful side effects. Unfortunately, the genetic technologies that are currently being developed and marketed do not fulfill these conditions at all.

Recent experimental trials have shown that GM seeds do not increase crop yields significantly.[83] Moreover, there are strong indications that the widespread use of GM crops will not only fail to solve the problem of hunger but, on the contrary, may perpetuate and even aggravate it. If transgenic seeds continue to be developed and promoted exclusively by private corporations, poor farmers will not be able to afford them, and if the biotech industry continues to protect its products by means of patents that prevent farmers from storing and trading seeds, the poor will become further dependent and marginalized. According to a recent report by the charitable organization Christian Aid, "GM crops are . . . creating classic preconditions for hunger and famine. Ownership of resources concentrated in too few hands—inherent in farming based on patented proprietary products—and a food supply based on too few varieties of crops widely planted are the worst option for food security."[84]

An Ecological Alternative

If the chemical and genetic technologies of our agroindustry will not alleviate world hunger but will continue to ruin the soil, perpetuate social injustice, and endanger the ecological balance in our natural environment, where can we turn to solve these problems? Fortunately, there is a well-documented and widely proven solution—a solution both time-honored and new that is now slowly sweeping the farming world in a quiet revolution. It is an ecological alternative, known variously as "organic farming," "sustainable agriculture," or "agroecology."[85]

When farmers grow crops "organically," they use technologies based on ecological knowledge rather than chemistry or genetic engineering to increase yields, control pests, and build soil fertility. They plant a variety of crops, rotating them so that insects that are attracted to one crop will disappear with the next. They know that it is unwise to eradicate pests completely, because this would also eliminate the natural predators that keep pests in balance in a healthy ecosystem. Instead of chemical fertilizers, these farmers enrich their fields with manure and tilled-in crop residue, thus returning organic matter to the soil to reenter the biological cycle.

Organic farming is sustainable because it embodies ecological principles that have been tested by evolution for billions of years.[86] Organic farmers know that a fertile soil is a living soil containing billions of living organisms in every cubic centimeter. It is a complex ecosystem in which the substances that are essential to life move in cycles from plants to animals, to manure, to soil bacteria, and back to plants. Solar energy is the natural fuel that drives these ecological cycles, and living organisms of all sizes are necessary to sustain the whole system and keep it in balance. Soil bacteria carry out various chemical transformations, such as the process of nitrogen fixation that makes atmospheric nitrogen accessible to plants. Deep-rooted weeds bring minerals to the soil surface where crops can make use of them. Earthworms break up the soil and loosen its texture; and all these activities are interdependent, combining to provide the nourishment that sustains life on Earth.

Organic farming preserves and sustains the great ecological cycles, integrating their biological processes into the processes of food production. When soil is cultivated organically, its carbon content increases, and thus organic farming contributes to reducing global warming. Physicist Amory Lovins estimates that increasing the carbon content of the world's depleted soils at plausible rates would absorb about as much carbon as all human activity emits.[87]

Animals are raised on organic farms to support the ecosystems above the ground and in the soil, and the whole enterprise is labor-intensive and community-oriented. Farms tend to be small and owner-operated. Their products are sold more at farmers' markets than in supermarkets, which shortens the distance "from the farm to the table," saving energy and packaging and maintaining the freshness of the food.[88]

The current renaissance in organic farming is a worldwide phenomenon. Farmers in over 130 countries now produce organic food commercially. The total area being farmed sustainably is estimated at more than 7 million hectares (17 million acres), and the market for organic food has grown to an estimated $22 billion a year.[89]

Scientists at a recent international conference on sustainable agriculture in Bellagio, Italy, reported that a series of large-scale experimental projects around the world that tested agroecological techniques—crop rotation, intercropping, use of mulches and compost, terracing, water harvesting, etc.—yielded spectacular results.[90] Many of these were achieved in resource-poor areas that had been deemed incapable of producing food surpluses. For example, agroecological projects involving about 730,000 farm households across Africa resulted in yield increases of between 50 and 100 percent, while decreasing production costs, increasing cash incomes of households dramatically—sometimes by as much as ten times. Again and again it was demonstrated that organic farming not only increased production and offered a wide range of ecological benefits, but also empowered the farmers. As one Zambian farmer put it, "Agroforestry has restored my dignity. My family is no longer hungry; I can even help my neighbours now."[91]

In southern Brazil, the use of cover crops to increase soil activity and water retention enabled 400,000 farmers to increase maize and soybean yields by over 60 percent. In the Andean region, increases in crop varieties resulted in twentyfold increases in yields and more. In Bangladesh, an integrated rice-fish program raised rice yields by 8 percent and farmers' incomes by 50 percent. In Sri Lanka, integrated pest and crop management increased rice yields by 11 to 44 percent while augmenting net incomes by 38 to 178 percent.

The Bellagio Report emphasizes that the innovative practices it documents involved whole communities and relied on existing local knowledge and resources as well as on scientific insight. Thus, "the new methods rapidly spread among farmers, which showed the potential for farmer-led dissemination of even complex technologies when users are actively engaged in understanding and adapting them instead of just being trained to use them."[92]

The Hazards of Genetic Engineering in Agriculture

There is now abundant evidence that organic farming is a sound ecological alternative to the chemical and genetic technologies of industrial agriculture. As Miguel Altieri concludes, organic farming "raise[s] agricultural productivity in economically viable, environmentally benign, and socially uplifting ways."[93] Unfortunately, none of that can be said about the current applications of genetic engineering to agriculture.

The risks of current biotechnologies in agriculture are a direct consequence of our poor understanding of genetic function. We have only recently come to realize that all biological processes involving genes are regulated by the cellular networks in which genomes are embedded, and that the patterns of genetic activity change continually in response to changes in the cellular environment. Biologists are only just beginning to shift their attention from genetic structures to metabolic networks, and they still know very little about the complex dynamics of these networks.

We also know that all plants are embedded in complex ecosystems, both above the ground and in the soil, in which inorganic and organic matter moves in continual cycles. Again, we know very little about these ecological cycles and networks—partly because for many decades the dominant genetic determinism resulted in a severe distortion of biological research, with most of the funding going into molecular biology and very little into ecology.

Since the cells and regulatory networks of plants are relatively simpler than those of animals, it is much easier for geneticists to insert foreign genes into plants. The problem is that once the foreign gene is in the plant's DNA and the resulting transgenic crop has been planted, it becomes part of an entire ecosystem. The scientists working for biotech companies know very little about the ensuing biological processes, and even less about the ecological consequences of their actions.

The most widespread use of plant biotechnology has been to develop herbicide-tolerant crops in order to boost the sales of particular herbicides. There is a strong likelihood that the transgenic plants will cross-pollinate with wild relatives in their surroundings, thus creating herbicide-resistant "superweeds." Evidence indicates that such gene flows between transgenic crops and wild relatives are already occurring.[94] Another serious problem is the risk of cross-pollination between transgenic crops and organically grown crops in nearby fields, which jeopardizes the organic farmers' important need to have their produce certified as truly organic.

To defend their practices, biotech supporters often claim that genetic engineering is like conventional breeding—a continuation of the age-old tradition of shuffling genes to obtain superior crops and livestock. Sometimes they even argue that our modern biotechnologies represent the latest stage in nature's adventure of evolution. Nothing could be farther from the truth. To begin with, the pace of gene alteration through biotechnology is several orders of magnitude faster than nature's. No ordinary plant breeder would be able to alter the genomes of half of the world's soybeans in just three years. Genetic modification of crops is undertaken with incredible haste, and transgenic crops are planted massively without proper advance testing of the short- and long-term impacts on ecosystems and human health. These untested and potentially hazardous GM crops are now spreading all over the world, creating irreversible risks.

A second difference between genetic engineering and conventional breeding is that conventional breeders transfer genes between varieties that interbreed naturally, whereas genetic engineering enables biologists to introduce a completely new and exotic gene into the genome of a plant—a gene from another plant or an animal with whom the plant can never mate naturally. Scientists cross the natural species barriers with the help of the aggressive gene transfer vectors, many of which are derived from disease-causing viruses that may recombine with existing viruses to create new pathogens.[95] As a biochemist put it at a recent conference: "Genetic engineering resembles more a viral infection than traditional breeding."[96]

The global fight for market share dictates not only the pace of production and deployment of transgenic crops, but also the direction of basic research. This is perhaps the most disturbing difference between genetic engineering and all previous shuffling of genes through evolution and natural breeding. In the words of the late biophysicist Donella Meadows: "Nature selects according to the ability to thrive and reproduce in the environment. Farmers have selected for 10,000 years according to what feeds people. Now the criterion is what can be patented and sold."[97]

Since one of the main objectives of plant biotechnology so far has been to increase the sales of chemicals, many of its ecological hazards are similar to those created by chemical agriculture.[98] The tendency to create broad international markets for a single product generates vast monocultures that reduce biodiversity, thus diminishing food security and increasing vulnerability to plant diseases, insect pests, and weeds. These problems are especially acute in developing countries, where traditional systems of diverse crops and foods are being replaced by monocultures that push countless species to extinction and create new health problems for rural populations.[99]

The story of the genetically engineered "golden rice" is a poignant example. A few years ago, a small team of idealistic geneticists without industry support created a yellow rice with high levels of beta-carotene, which is converted to vitamin A in the human body. The rice was promoted as a cure for the blindness and vision impairment caused by vitamin A deficiency. According to the UN, vitamin A deficiency currently affects more than 2 million children.

The news of this "miracle cure" was received enthusiastically by the press, but closer examination has shown that instead of helping children at risk, the project is likely to repeat the mistakes of the Green Revolution while adding new hazards for ecosystems and human health.[100] By reducing biodiversity, cultivation of vitamin A rice will eclipse alternative sources of vitamin A that are available in traditional agricultural systems. Agroecologist Vandana Shiva points out that women farmers in Bengal, for example, use numerous varieties of green leafy vegetables that are an excellent source of beta-carotene. Those who suffer the highest rates of vitamin A deficiency are the poor, who suffer from malnutrition in general and who would benefit much more from the development of sustainable, community-based agriculture than from GM crops they cannot afford.

In Asia, vitamin A from native greens and fruits is often produced without irrigation, whereas the cultivation of rice is water-intensive and would require the mining of ground water or the construction of large dams, with all of their associated environmental problems. Moreover, as in the case of other GM crops, we still know very little about the ecological impact of vitamin A rice on soil organisms and other rice-dependent species in the food chain. "Promoting it as a tool against blindness while ignoring safer, cheaper, available alternatives provided by our rich agrobiodiversity," Shiva concludes, "is nothing short of a blind approach to blindness control."

Most of the ecological hazards associated with herbicide-resistant crops, such as Monsanto's Roundup Ready soybeans, derive from the ever-increasing use of the company's herbicide. Since resistance to that specific herbicide is the crop's only—and widely advertised—benefit, farmers are naturally led to use massive amounts of the weed-killer. It is well documented that such massive use of a single chemical greatly boosts herbicide resistance in weed populations, which triggers a vicious cycle of more and more intensive spraying.

Such use of toxic chemicals in agriculture is especially harmful to consumers. When plants are sprayed repeatedly with a weed-killer, they retain chemical residues that show up in our food. Moreover, plants grown in the presence of massive amounts of herbicides can suffer from stress and will typically respond by over- or underproducing certain substances. Herbicide-resistant members of the bean family are known to produce higher levels of plant oestrogens, which may cause severe dysfunctions in human reproductive systems, especially in boys.[101]

Almost 80 percent of today's acreage of GM crops is planted with herbicide-resistant varieties. The remaining 20 percent consists of so-called "insect-resistant" crops. These are genetically engineered to produce pesticides in every one of their cells throughout their entire life cycles. The best-known example is a naturally occurring insecticide, a bacterium called *Bacillus thuringiensis* and commonly known as Bt, whose toxin-producing genes have been spliced into cotton, corn, potato, apple, and several other plants.

The resulting transgenic crops are immune to some insects. However, since most crops are subjected to a diversity of insect pests, insecticides still have to be applied. A recent U.S. study found that at seven out of twelve sites there was no significant difference in pesticide use between Bt crops and non-Bt crops. In one site, the use of pesticides on Bt cotton was even higher than on non-Bt cotton.[102]

The ecological hazards of Bt crops are a consequence of important differences between the naturally occurring Bt bacteria and genetically modified Bt crops. Organic farmers have used the Bt bacterium as a natural pesticide for over fifty years to control leaf-eating caterpillars, beetles, and moths. They use it judiciously, dusting their crops only occasionally so that insects are not able to develop resistance. But when Bt is produced continuously inside crops that are planted over hundreds of thousands of acres, their pests are constantly exposed to the toxin and will inevitably become resistant to it.

Consequently, Bt will rapidly become useless both in GM crops and as a natural pesticide. Plant biotechnology will have destroyed one of the most important biological tools of integrated pest management. Even scientists in the biotech industry acknowledge that Bt will become useless within ten years, but the biotech companies seem to calculate cynically that their patents on Bt technology will have run out by then and they will have moved on to create other types of insecticide-producing plants.

Another difference between natural Bt and Bt-producing crops is that the latter seem to harm a wider range of insects, including many that are beneficial to the ecosystem as a whole. In 1999, a study published in *Nature* about caterpillars of the Monarch butterfly being killed by pollen from Bt corn attracted widespread public attention.[103] Since then it has been found that Bt toxins from GM crops also affect ladybugs, bees, and other beneficial insects.

The Bt toxins in GM plants are also harmful to soil ecosystems. As farmers incorporate crop residues into the ground after harvest, the toxins accumulate in the soil, where they may cause serious harm to the myriads of microorganisms that make up a healthy soil ecosystem.[104]

In addition to the harmful effects of Bt crops on ecosystems above and below the ground, the direct hazards to human health are obviously a major concern. At present, we know very little about the potential effects of these toxins on the microorganisms that are vital to our digestive system. However, since numerous effects on soil microbes have already been observed, we should be concerned about the pervasive presence of Bt toxins in corn, potatoes, and other food crops.

The environmental risks of current plant biotechnologies are evident to any agroecologist, even though the detailed effects of GM crops on agricultural ecosystems are still poorly understood. In addition to these expected risks, numerous unexpected side effects have been observed in genetically modified plant and animal species.[105]

Monsanto is now facing an increasing number of lawsuits from farmers who had to cope with these unexpected side effects. For example, the balls of their GM cotton were deformed and dropped off in thousands of acres in the Mississippi Delta; their GM canola seeds had to be pulled off the Canadian market because of contamination with a hazardous gene. Similarly, Calgene's Flavr-Savr tomato, engineered for improved shelf life, was a commercial disaster and soon disappeared. Transgenic potatoes intended for human consumption caused a series of serious health problems when they were fed to rats, including tumor growth, liver atrophy, and shrinkage of the brain.[106]

In the animal kingdom, where cellular complexity is much higher, the side effects in genetically modified species are much worse. "Supersalmon," which were engineered to grow as fast as possible, ended up with monstrous heads and died from not being able to breathe or feed properly. Similarly, a "superpig" with a human gene for a growth hormone turned out ulcerous, blind, and impotent.

The most horrifying and by now best-known story is probably that of the genetically altered hormone called "recombinant bovine growth hormone," which has been used to stimulate milk production in cows despite the fact that American dairy farmers have produced vastly more milk than people can consume for the past fifty years. The effects of this genetic engineering folly on the cows' health are serious. They include bloat, diarrhea, diseases of the knees and feet, cystic ovaries, and many more. Besides, their milk may contain a substance that has been implicated in human breast and stomach cancers.

Because these GM cows require more protein in their diet, their feed was supplemented with ground-up animals in some countries. This completely unnatural practice, which turns cows from vegetarians into cannibals, has been associated with the recent epidemic of BSE ("mad cow disease") and increased incidence of its human analogue, variant Creutzfeldt-Jakob disease. This is one of the most extreme cases of biotechnology gone haywire. As biologist David Ehrenfeld points out, "There seems little reason to increase the risk of this terrible disease for the sake of a biotechnology that we don't need. If cows stay off hormones and concentrate on eating grass, all of us will be much better off."[107]

As genetically modified foods begin to flood our markets, their health risks are aggravated by the fact that the biotech industry, with support from government regulatory agencies, refuses to label them properly, so that consumers cannot discriminate between GM and non-GM food. In the United States, the biotech industry has persuaded the Food and Drug Administration (FDA) to treat GM food as "substantially equivalent" to traditional food, which allows food producers to evade normal testing by the FDA and the Environmental Protection Agency (EPA), and also leaves it to the companies' own discretion as to whether to label their products as genetically modified. Thus, the public is kept unaware of the rapid spread of transgenic foods and scientists will find it much harder to trace harmful effects. Indeed, buying organic is now the only way to avoid GM foods.

Confidential documents made public in a class-action lawsuit have revealed that even scientists within the FDA do not agree with the concept of "substantial equivalence."[108] Besides, the position of the biotech industry contains an inherent contradiction. On the one hand, the industry claims that its crops are substantially equivalent to traditional crops and hence do not need to be labeled, or tested; on the other hand, it insists that they are novel and therefore can be patented. As Vandana Shiva sums it up, "A myth of 'substantial equivalence' has been created to deny citizens the right to safety and deny scientists the right to practise sound and honest science."[109]

Life as the Ultimate Commodity

In their attempts to patent, exploit, and monopolize all aspects of biotechnology, the top agrochemical corporations have bought up seed and biotech companies and have restyled themselves as "life sciences corporations."[110] Traditional boundaries between pharmaceutical, agrochemical, and biotechnology industries are rapidly disappearing as corporations merge to form giant conglomerates under the life sciences banner. Thus Ciba-Geigy merged with Sandoz to become Novartis; Hoechst and Rhone Poulenc became Aventis; and Monsanto now owns and controls several large seed companies.

What all these life sciences corporations have in common is a narrow understanding of life, based on the erroneous belief that nature can be subjected to human control. This ignores the self-generating and self-organizing dynamic that is the very essence of life and instead redefines living organisms as machines that can be managed from outside and be patented and sold as industrial resources. Life itself has become the ultimate commodity.

As Vandana Shiva reminds us, the Latin root of the word "resource" is *resurgere* ("to rise again"). In the ancient meaning of the term, a natural resource, like all of life, is inherently self-renewing. This profound understanding of life is denied by the new life sciences corporations when they prevent life's self-renewal in order to turn natural resources into profitable raw materials for industry. They do so through a combination of genetic alterations (including the terminator technologies)[111] and patents, which do violence to time-honored farming practices that respect the cycles of life.

Since a patent is traditionally understood as the exclusive right to the use and selling of an invention, it seems strange that biotech companies today are able to patent living organisms, from bacteria to human cells. The history of this achievement is an amazing story of scientific and legal sleight of hand.[112] The patenting of life-forms became common practice in the 1960s when property rights were given to plant breeders for new varieties of flowers obtained through human intervention and ingenuity. It took the international legal community less than twenty years to move from this seemingly harmless patenting of flowers to the monopolization of life.

Next, specially bred food plants were patented, and soon after lawmakers and regulators argued that there was no theoretical basis for preventing the extension of industrial patenting from plants to animals and microorganisms. Indeed, in 1980 the U.S. Supreme Court handed down the landmark decision that genetically modified microorganisms could be patented.

What was conveniently ignored in these legal arguments was the fact that the original plant patents for *improved* varieties of flowers did not extend to the source material, which was considered the "common heritage of mankind."[113] The patents now granted to biotech companies, by contrast, cover not only the methods by which DNA sequences are isolated, identified, and transferred, but also the underlying genetic material itself. Moreover, existing national laws and international conventions that specifically prohibit the patenting of essential natural resources, such as food and plant-derived medicine, are now being altered in accordance with the corporate view of life as a profitable commodity.

In recent years, the patenting of life forms has given rise to a new form of "biopiracy." Gene hunters prospect countries in the South for valuable genetic resources, such as seeds of special crops or medicinal plants, often with the help of indigenous communities who trustingly provide the materials together with their knowledge. These resources are then taken to biotech laboratories in the North, where they are isolated, genetically identified, and patented.[114]

These exploitative practices are legalized by the WTO's narrow definition of intellectual property rights (IPRs), which recognizes knowledge as patentable only when it is expressed within the framework of Western science. As Vandana Shiva points out, "This excludes all kinds of knowledge, ideas, and innovations that take place in the intellectual commons—in villages among farmers, in forests among tribespeople, and even in universities among scientists."[115] Thus the exploitation of life is extended even beyond living organisms to the knowledge and collective innovations of indigenous communities. "With no regard or respect for other species and cultures," Shiva concludes, "IPRs are a moral, ecological, and cultural outrage."

The Turning of the Tide

In recent years, the health problems caused by genetic engineering, as well as its deeper social, ecological, and ethical problems, have become all too apparent, and there is now a rapidly growing global movement that rejects this form of technology.[116] Numerous health and environmental organizations have called for a moratorium on commercial releases of genetically modified organisms pending a comprehensive public inquiry into the legitimate and safe uses of genetic engineering.[117] These appeals also propose that there should be no patents on living organisms or their parts, and that the basis of our approach to biotechnology should be the precautionary principle, which has been written into international agreements since the Earth Summit in 1992. Technically known as Principle 15 of the Rio declaration, it states: "Where there are threats of serious or irreversible damage, lack of full scientific certainty shall not be used as a reason for postponing cost-effective measures to prevent environmental degradation."

The shift of emphasis in molecular biology from the structure of genetic sequences to the organization of genetic and epigenetic networks, from genetic programs to emergent properties also means that calls for a radically new approach to biotechnology are coming forth not only from ecologists, health professionals, and concerned citizens, but increasingly from leading geneticists, as I have documented throughout this chapter. With the intriguing discoveries of the Human Genome Project, the discussion of the current paradigm shift in biology has now even reached the popular scientific press. It is significant, in my view, that a special science section of the *New York Times* on the results of the Human Genome Project pictured the genome for the first time as a complex functional network (see opposite page).

Once the systems view of life has been embraced by our scientists, engineers, and political and corporate leaders, we can imagine a radi-

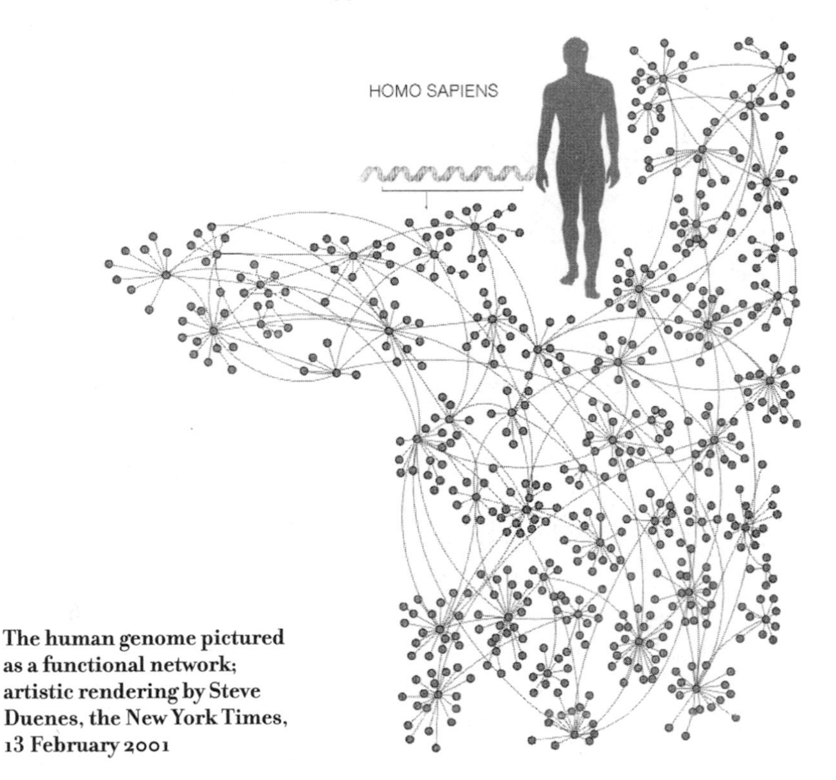

The human genome pictured as a functional network; artistic rendering by Steve Duenes, the New York Times, 13 February 2001

cally different kind of biotechnology. It would start with the desire to learn from nature rather than control her, using nature as a mentor rather than merely as a source of raw materials. Instead of treating the web of life as a commodity, we would respect it as the context of our existence.

This new type of biotechnology would not involve modifying living organisms genetically but instead would use the techniques of genetic engineering to understand nature's subtle "designs" and use them as models for new human technologies. We would integrate ecological knowledge into the design of materials and technological processes, learning from plants, animals, and microorganisms how to manufacture fibers, plastics, and chemicals that are nontoxic, completely biodegradable, and subject to continual recycling.

These would be *bio*technologies in a new sense, because life's material structures are based on proteins that we could produce only with the help of enzymes supplied by living organisms. The development of such new biotechnologies will be a tremendous intellectual challenge, because we still do not understand how nature developed "technologies" during billions of years of evolution that are far superior to our human designs. How do mussels produce glue that sticks to anything in water? How do spiders spin a silk thread that, ounce for ounce, is five times stronger than steel? How do abalone grow a shell that is twice as tough as our high-tech ceramics? How do these creatures manufacture their miracle materials in water, at room temperature, silently, and without any toxic byproducts?

To find the answers to these questions and use them to develop technologies inspired by nature could provide fascinating research programs for scientists and engineers for decades to come. Indeed, these programs have already begun. They are part of an exciting new field of engineering and design known as "biomimicry" and, more generally, as "ecodesign," which has recently generated a burst of optimism about humanity's chances of moving toward a sustainable future.[118]

In her book *Biomimicry*, science writer Janine Benyus takes us on a fascinating journey to numerous laboratories and field stations where interdisciplinary teams of scientists and engineers analyze the detailed chemistry and molecular structures of nature's most complex materials to use them as models for new biotechnologies.[119] They are discovering that many of our major technological problems have already been solved in nature in elegant, efficient, and ecologically sustainable ways, and they are trying to adapt these solutions for human use.

Scientists at the University of Washington have studied the molecular structure and assembly process of the smooth inner coating of abalone shells, which shows delicate swirling color patterns and is hard as nails. They were able to mimic the assembly process at ambient temperatures and create a hard, transparent material that could be an ideal coating for the windshields of ultralight electric cars. German researchers have mimicked the bumpy, self-cleaning micro-surface of the lotus leaf to produce a paint that will do the same for buildings. Marine biologists and biochemists have spent many years analyzing the unique chemistry used by blue mussels to secrete an adhesive that bonds underwater. They are now exploring potential medical applications that would allow surgeons to create bonds between ligaments and tissues in a fluid environment. Physicists have teamed up with biochemists in several laboratories to examine the complex structures and processes of photosynthesis, eventually hoping to mimic them in new kinds of solar cells.

While these exciting developments are taking place, however, the central assertion of genetic determinism that genes determine behavior is still perpetuated by many geneticists, in biotechnology companies as well as in the academic world. One has to wonder whether these scientists really believe that our behavior is determined by our genes, and if not, why they keep up this façade.

Discussions of this issue with molecular biologists have shown me that there are several reasons why scientists feel that they have to perpetuate the dogma of genetic determinism in spite of mounting contrary evidence. Industrial scientists are often hired for specific, narrowly defined projects, work under strict supervision, and are forbidden to discuss the broader implications of their research. They are required to sign confidentiality clauses to that effect. In biotechnology companies, in particular, the pressure to conform with the official doctrine of genetic determinism is enormous.

In the academic world the pressures are different but, unfortunately, almost equally strong. Because of the tremendous cost of genetic research, biology departments increasingly form partnerships with biotechnology corporations to receive substantial grants that shape the nature and direction of their research. As Richard Strohman observes, "Academic biologists and corporate researchers have become indistinguishable, and special awards are now given for collaborations between these two sectors for behaviour that used to be cited as a conflict of interest."[120]

Biologists are used to formulating their grant proposals in terms of genetic determinism, because they know that this is what gets funded. They promise their funders that certain results will be derived from the future knowledge of genetic structure even though they know well that scientific advances are always unexpected and unpredictable. They learn to adopt this double standard during their years as graduate students and then keep it up throughout their academic careers.

In addition to these evident pressures, there are more subtle cognitive and psychological barriers that prevent biologists from embracing the systems view of life. Reductionism is still the dominant paradigm in their education, and hence they are often unfamiliar with concepts like self-organization, networks, or emergent properties. Besides, genetic research even within the reductionist paradigm can be tremendously exciting: the mapping of genomes is an amazing achievement that would have been unthinkable for scientists a mere generation ago. It is quite understandable that many geneticists get carried away and want to continue their well-funded research without worrying about the broader implications.

Finally, we need to remember that science is an intensely collective enterprise. Scientists feel a great need to belong to their intellectual communities and will not easily speak out against them. Even tenured scientists who have had brilliant careers and received prestigious awards are often reluctant to raise a critical voice.

In spite of these barriers the worldwide opposition to the patenting, marketing, and release of genetically modified organisms, combined with the recently revealed limitations of the conceptual foundations of genetic engineering, show that the edifice of genetic determinism is now crumbling. To quote Evelyn Fox Keller once more, "It seems evident that the primacy of the gene as the core explanatory concept of biological structure and function is more a feature of the twentieth century than it will be of the twenty-first."[121] In conclusion, it is becoming increasingly apparent that biotechnology is now reaching a scientific, philosophical, and political turning point.

| seven |

CHANGING THE GAME

As this new century unfolds, it becomes increasingly apparent that the neoliberal "Washington Consensus" and the policies and economic rules set forth by the Group of Seven and their financial institutions—the World Bank, the IMF, and the WTO—are consistently misguided. Analyses by scholars and community leaders cited throughout this book show that the "new economy" is producing a multitude of interconnected harmful consequences—rising social inequality and social exclusion, a breakdown of democracy, more rapid and extensive deterioration of the natural environment, and increasing poverty and alienation. The new global capitalism has also created a global criminal economy that profoundly affects national and international economies and politics; it has threatened and destroyed local communities around the world; and with the pursuit of an ill-conceived biotechnology it has invaded the sanctity of life by attempting to turn diversity into monoculture, ecology into engineering, and life itself into a commodity.

State of the World

Despite new environmental regulations, the increasing availability of ecofriendly products and many other encouraging developments championed by the environmental movement, the massive loss of forests and the greatest extinction of species in millions of years has not been reversed.[1] By depleting our natural resources and reducing the planet's biodiversity we damage the very fabric of life on which our well-being depends, including the priceless "ecosystem services" that nature provides for free—processing waste, regulating the climate, regenerating the atmosphere, and so on.[2] These vital processes are emergent properties of nonlinear living systems that we are only beginning to understand, and they are now seriously endangered by our linear pursuits of economic growth and material consumption.

These dangers are exacerbated by the global climate change produced by our industrial systems. The causal link between global warming and human activity is no longer hypothetical. In late 2000, the authoritative Intergovernmental Panel on Climate Change (IPCC) published its strongest consensus statement to date that human release of carbon dioxide and other greenhouse gases "contributed significantly to the observed warming over the last fifty years."[3] By the end of the century, the IPCC predicted, temperatures could soar by almost 6°C. This would be an increase exceeding the change of temperature between the last Ice Age and today. As a consequence, virtually every natural system on Earth and every human economic system would be at risk from rising water levels, more severe storms, and more intense droughts.[4]

Although there have recently been some declines in global carbon emissions, they have failed to slow the rate of global climate change. On the contrary, recent evidence indicates that it is accelerating. This evidence comes from two separate and equally troubling observations—the rapid melting of glaciers and Arctic Sea ice, and the declining health of coral reefs.

The melting of glaciers at extraordinary rates around the world is one of the most ominous signs of the warming caused by the continuing reckless burning of fossil fuels. Moreover, in July 2000, scientists who reached the North Pole aboard the Russian icebreaker *Yamal* were confronted with a strange and eerie sight—an expanse of open water, about a mile wide, in place of the thick ice that has for ages covered the Arctic Ocean.[5]

If this massive melting continues, it will have dramatic global effects. Arctic ice is an important element in the dynamics of the Gulf Stream, as scientists have recently learned. Removing it from the North Atlantic circulation system would drastically change Europe's climate and affect other parts of the world.[6] Moreover, the diminished ice cover would reflect less sunlight and hence would further accelerate the Earth's warming, setting in motion a vicious cycle. In the worst-case scenario of IPCC's scientists, the snows of Kilimanjaro, immortalized in Hemingway's famous short story, could disappear within fifteen years; and so could the snow in the Alps.

Less visible than the melting of glaciers in the high mountains, but equally significant, is alarming evidence of increased global warming from the tropical oceans. In many parts of the tropics, shallow waters house huge coral reefs that were built by tiny polyps over long periods of geological time. These massive structures—by far the largest created by living organisms on Earth—support innumerable plants, animals, and microorganisms. Aside from the tropical rainforests, the tropical coral reefs are the most complex ecosystems on Earth, true wonders of biodiversity.[7]

In recent years, coral reefs around the world, from the Caribbean to the Indian Ocean and Australia's Great Barrier Reef, have experienced life-threatening environmental stresses, partly due to rising temperatures. Coral polyps are extremely sensitive to temperature changes and may turn white and die when the ocean temperature rises even slightly. In 1998, marine biologists estimated that more than one quarter of the world's coral reefs were sick or dying, and two years later, scientists reported that half of the vast coral reefs surrounding the Indonesian archipelago have been destroyed by the effects of marine pollution, deforestation, and rising temperatures.[8] This worldwide decimation of coral reefs is one of the clearest and most troubling indications that our planet is warming.

While scientists record telltale signs of global warming in the Arctic and in the tropics, "natural" disasters with devastating effects that are caused in part by human-induced global climate change and other ecologically destructive practices are increasing in frequency. In 1998 alone, three such disasters struck in different parts of the world, each resulting in the loss of thousands of human lives and exacting catastrophic financial tolls.[9]

Hurricane Mitch, the deadliest Atlantic storm in 200 years, took 10,000 lives and devastated large areas of Central America, setting back development in the region by decades. The effects of the storm were aggravated by the interplay of climate change, deforestation due to population pressures, and soil erosion. In China, the catastrophic flood of the Yangtze River, which caused more than 4,000 deaths and inundated 25 million hectares (62 million acres) of cropland, was largely a consequence of deforestation that had left many steep hillsides bare. In the same year, Bangladesh suffered its most devastating flood of the century, which killed 1,400 people and inundated two-thirds of the country for several months. The flood was exacerbated by rain falling on heavily logged areas and by runoffs from extensive developments upstream clogging the region's rivers.

Sea levels are rising steadily due to global warming. They rose about 20 centimeters during the last century and, if current trends continue, will rise another 50 centimeters by 2100. Meteorologists predict that this would put the world's major river deltas—Bangladesh, the Amazon, and the Mississippi—at risk, and that rising sea levels could even flood the New York City subway system.[10]

The (often literally) rising tide of natural catastrophes over the past decade is a clear indication that the climatic instability caused by human actions is increasing, while we are also disrupting the services of healthy ecosystems that provide protection from natural disasters. As the Worldwatch Institute's Janet Abramovitz points out:

> Many ecosystems have been frayed to the point where they are no longer resilient and able to withstand natural disturbances, setting the stage for "unnatural disasters"—those made more frequent or more severe due to human actions. By destroying forests, damming rivers, filling in wetlands, and destabilizing the climate, we are unravelling the strands of a complex ecological safety net.[11]

Careful analysis of the dynamics underlying recent natural disasters also shows that environmental and social stresses are tightly interconnected in all of them.[12] Poverty, scarcity of resources, and expanding populations combine to create vicious cycles of degradation and breakdown in both ecosystems and local communities.

The principal lesson to be learned from these analyses is that the causes of most of our present environmental and social problems are deeply embedded in our economic systems. As I emphasized previously, the current form of global capitalism is ecologically and socially unsustainable, and hence politically unviable in the long run.[13] More stringent environmental regulations, better business practices, and more efficient technologies are all necessary, but they are not enough. We need a deeper systemic change.

Such deep systemic change is already under way. Scholars, community leaders, and grassroots activists around the world are forming effective coalitions and are raising their voices not only to demand that we must "change the game," but also to suggest concrete ways of doing so.

Globalization by Design

Any realistic discussion of changing the game must begin with the recognition that, although globalization is an emergent phenomenon, the current form of economic globalization has been consciously designed and *can* be reshaped. As we have seen, today's global economy is structured around networks of financial flows in which capital works in real time, moving rapidly from one option to another in a relentless search for investment opportunities.[14] The global market is really a network of machines—an automaton that imposes its logic on all human participants. However, in order to function smoothly, this automaton has to be programmed by human actors and institutions. The programs that give rise to the new economy consist of two essential components—values and operational rules.

The global financial networks process signals that assign a specific financial value to every asset in every economy. This process is far from straightforward. It involves economic calculations based on advanced mathematical models; information and opinions provided by market valuation firms, financial gurus, leading central bankers, and other influential analysts; and, last but not least, information turbulences that are largely uncontrolled.[15]

In other words, the tradable financial value of any asset (which is subject to continual adjustments) is an emergent property of the automaton's highly nonlinear dynamics. However, underlying all evaluations is the basic principle of unfettered capitalism: that moneymaking should always be valued higher than democracy, human rights, environmental protection, or any other value. Changing the game means, first and foremost, changing this basic principle.

In addition to the complex process of assessing tradable values, the programs of the global financial networks contain operational rules that must be followed by markets around the world. These are the free-trade rules that the World Trade Organization (WTO) imposes on its member states. To ensure maximum profit margins in the global casino, capital must be allowed to flow freely through its financial networks so that it can be invested anywhere in the world at a moment's notice. These free-trade rules, together with increasing deregulation of corporate activities, are designed to guarantee the free movement of capital. The impediments to unrestricted trade that are removed or curtailed by this new legal framework are usually environmental regulations, public health laws, food safety laws, workers' rights, and laws giving nations control over investments on their territory and ownership of their local culture.[16]

The resulting integration of economic activities goes beyond purely economic aspects: it extends to the cultural domain. Countries around the world with vastly different cultural traditions are increasingly homogenized through relentless proliferation of the same restaurant franchises, hotel chains, high-rise architecture, superstores, and shopping malls. The result, in Vandana Shiva's apt phrase, is an increasing "monoculture of the mind."

The economic rules of global capitalism are enforced and vigorously promoted by three global financial institutions—the World Bank, the IMF, and the WTO. They are known collectively as the Bretton Woods institutions because they were established at a UN conference in Bretton Woods, New Hampshire, in 1944, in order to create an institutional framework for a coherent worldwide postwar economy.

The World Bank was originally created to finance the postwar reconstruction of Europe, and the IMF to assure the stability of the international financial system. However, both institutions soon shifted their focus to promoting and enforcing a narrow model of economic development in the Third World, often with disastrous social and environmental consequences.[17] The ostensible role of the WTO is to regulate trade, prevent trade wars and protect the interests of poor nations. In reality, the WTO implements and enforces globally the same agenda that the World Bank and the IMF have imposed on most of the developing world. Rather than protecting people's health, safety, livelihood, and culture, the WTO's free-trade rules undermine these basic human rights in order to consolidate the power and wealth of a small corporate elite.

The free-trade rules are the result of many years of negotiations behind closed doors, which involved industry trade groups and corporations, but excluded nongovernmental organizations (NGOs) representing the interests of the environment, social justice, human rights, and democracy. Not surprisingly, the worldwide anti-WTO movement is now demanding greater transparency in the establishment of market rules and independent reviews of the ensuing social and environmental consequences. A powerful coalition of hundreds of NGOs is now proposing a whole new set of trade policies that would profoundly change the global financial game.

Community leaders and grassroots movements around the world, social scientists, and even some of the most successful financial speculators are now beginning to agree that global capitalism needs to be regulated and constrained, and that its financial flows need to be organized according to different values.[18] At the 2001 meeting of the World Economic Forum in Davos, the exclusive club of representatives from big business, some of the leading players admitted for the first time that globalization has no future unless it is designed to be inclusive, ecologically sustainable, and respectful of human rights and values.[19]

There is a huge difference between making politically correct statements and actually changing corporate behavior, but agreeing on the basic values that are needed to reshape globalization would be a critical first step. What are these basic values? To reiterate Václav Havel's framing of the question, what are the ethical dimensions of globalization?[20]

Ethics refers to a standard of human conduct that flows from a sense of belonging. When we belong to a community, we behave accordingly.[21] In the context of globalization, there are two relevant communities to which we all belong. We are all members of humanity, and we all belong to the global biosphere. We are members of *oikos*, the "Earth household," which is the Greek root of the word "ecology," and as such we should behave as the other members of the household behave—the plants, animals, and microorganisms that form the vast network of relationships that we call the web of life.

This global living network has unfolded, evolved, and diversified for the last three billion years without ever being broken. The outstanding characteristic of the Earth household is its inherent ability to sustain life. As members of the global community of living beings, it behoves us to behave in such a way that we do not interfere with this inherent ability: this is the essential meaning of ecological sustainability. What is sustained in a sustainable community is not economic growth or development, but the entire web of life on which our long-term survival depends. It is designed so that its ways of life, businesses, economy, physical structures, and technologies do not interfere with nature's inherent ability to sustain life.

As members of the human community, our behavior should reflect a respect of human dignity and basic human rights. Since human life encompasses biological, cognitive, and social dimensions, human rights should be respected in all three of these dimensions. The biological dimension includes the right to a healthy environment and to secure and healthy food; honoring the integrity of life also includes the rejection of the patenting of life-forms. Human rights in the cognitive dimension include the right of access to education and knowledge, as well as the freedom of opinion and expression. In the social dimension, finally, the first human right—in the words of the UN Declaration of Human Rights—is "the right to life, liberty, and security of person." There is a wide range of human rights in the social dimension—from social justice to the right of peaceful assembly, cultural integrity, and self-determination.

In order to combine respect for these human rights with the ethics of ecological sustainability, we need to realize that sustainability—in ecosystems as well as in human society—is not an individual property but a property of an entire web of relationships: it involves a whole community. A sustainable human community interacts with other living systems—human and nonhuman—in ways that enable those systems to live and develop according to their nature. In the human realm sustainability is fully consistent with the respect of cultural integrity, cultural diversity, and the basic right of communities to self-determination and self-organization.

The Seattle Coalition

The values of human dignity and ecological sustainability, as outlined above, form the ethical basis for reshaping globalization, and an impressive global coalition of NGOs has formed around these values. The numbers of international nongovernmental organizations increased dramatically over the past few decades, from several hundred in the 1960s to over 20,000 by the end of the century.[22] During the 1990s, a computer-literate elite emerged within these international NGOs. They began to use new communications technologies very skillfully, especially the Internet, to network with one another, exchange information, and mobilize their members.

This networking became especially intense as they prepared joint protest actions for the meeting of the WTO in Seattle in November 1999. For many months, hundreds of NGOs interlinked electronically to coordinate their plans and to issue a flurry of pamphlets, position papers, press releases, and books in which they clearly articulated their opposition to the WTO's policies and undemocratic regime.[23] This literature was virtually ignored by the WTO, but had a significant impact on public opinion. The NGOS' educational campaign culminated in a two-day teach-in in Seattle before the WTO meeting, organized by the International Forum on Globalization and attended by over 2,500 people from around the world.[24]

On 30 November 1999, around 50,000 people belonging to more than 700 organizations took part in a superbly coordinated, passionate and almost entirely nonviolent protest that permanently changed the political landscape of globalization. As environmentalist and author Paul Hawken, who was one of the participants, saw it:

> No charismatic leader led. No religious figure engaged in direct action. No movie star starred. There was no alpha group. The Ruckus Society, Rainforest Action Network, Global Exchange, and hundreds more were there, co-ordinated primarily by cell phones, e-mails, and the Direct Action Network . . .
>
> They were organized, educated, and determined. They were human rights activists, labour activists, indigenous people, people of faith, steel workers, and farmers. They were forest activists, environmentalists, social justice workers, students, and teachers. And they wanted the World Trade Organization to listen. They were speaking on behalf of a world that has not been made better by globalization.[25]

The Seattle police turned out in force to keep the protesters away from the Convention Center where the meeting took place, but they were unprepared for the street actions of a massive, well-organized network totally committed to shutting down the WTO. Chaos ensued; hundreds of delegates were blocked off in the streets or confined to their hotels; and the opening ceremony had to be canceled.

The frustration of the delegates and politicians mounted as the day wore on. By late afternoon the mayor and police chief declared a state of civil emergency; and on the second day the police seemingly lost all control, brutally attacking not only protesters but also bystanders, commuters, and residents. Michael Meacher, Minister of the Environment of the U.K., stated that: "What we hadn't reckoned with was the Seattle Police Department, who single-handedly managed to turn a peaceful protest into a riot."[26]

Among the 50,000 demonstrators, there were perhaps 100 anarchists who had come to smash shop windows and destroy property. They could easily have been arrested, but the Seattle police neglected to do so, and the media chose to focus inordinately on the destructive actions of that tiny group of protesters—a fraction of 1 percent—rather than on the constructive message of the vast majority of nonviolent activists.

In the end, the WTO meeting broke down not only because of these massive demonstrations, but also—and perhaps even more so—because of the way the major powers within the WTO bullied the delegates from the south.[27] After ignoring dozens of proposals from developing countries, the WTO leaders excluded the delegates representing these countries from critical behind-the-scenes "Green Room" meetings and then pressured them to sign a secretly negotiated agreement. Infuriated, many developing countries refused to do so, thereby joining the massive opposition to the WTO's undemocratic regime that was going on outside the Convention Center.

Faced with the prospect of rejection by developing nations in the final session, the major powers preferred to let the Seattle meeting collapse without even attempting to issue a final declaration. Thus

Seattle, which was meant to be a celebration of the WTO's solidification, instead became the symbol of worldwide resistance.

After Seattle, smaller but equally effective demonstrations took place at other international meetings in Washington, Prague, and Quebec City, but Seattle was the turning point in the formation of a global coalition of NGOs. By the end of 2000, over 700 organizations from seventy-nine countries had joined what they now officially call the International Seattle Coalition, and began to launch a "WTO turnaround campaign."[28] Naturally, there is a great diversity of interests in these NGOs, which range from labor organizations to human rights, women's rights, religious, environmental, and indigenous peoples' organizations. However, there is remarkable agreement among them about the core values of human dignity and ecological sustainability.

In January 2001, the Seattle Coalition held the first World Social Forum in Porto Alegre, Brazil. Designed as a counterpoint to the World Economic Forum in Davos, Switzerland, it was intentionally held at the same time, but in the Southern hemisphere. The contrast between the simultaneous events was stark. In Switzerland, a small elite of mostly white and mostly male business leaders gathered in seclusion, protected from demonstrators by a huge contingent of the Swiss army. In Brazil, 12,000 women and men of all races met openly in vast lecture halls, warmly welcomed by the city of Porto Alegre and the entire state of Rio Grande do Sul.

For the first time, the Seattle Coalition had called its members together not to protest but to take the next step and discuss alternative scenarios, in keeping with the Forum's official motto, "Another World Is Possible." As the *Guardian* reported, "There was a tangible sense of an emerging global movement with a striking diversity of age, political traditions, practical experience and cultural background."[29]

Global Civil Society

The Seattle Coalition exemplifies a new kind of political movement that is typical of our Information Age. Skillful use of the Internet's interactivity, immediacy, and global reach means that NGOs in the coalition are able to network with each other, share information, and mobilize their members with unprecedented speed. As a result, the new global NGOs have emerged as effective political actors who are independent of traditional national or international institutions.

As we have seen, the rise of the network society has gone hand in hand with the decline of the sovereignty, authority, and legitimacy of the nation-state.[30] At the same time, mainstream religions have not developed an ethic appropriate for the age of globalization, while the legitimacy of the traditional patriarchal family is being challenged by profound redefinitions of gender relationships, family, and sexuality—the main institutions of traditional civil society are breaking down.

Civil society is traditionally defined as a set of organizations and institutions—churches, political parties, unions, cooperatives, and various voluntary associations—that form an interface between the state and its citizens. The institutions of civil society represent the interests of the people and constitute the political channels that connect them to the state. According to sociologist Manuel Castells, social change in the network society does not originate within the traditional institutions of civil society but develops from identities based on the rejection of society's dominant values—patriarchy, the domination and control of nature, unlimited economic growth and material consumption, and so on.[31] The resistance against these values originated in the powerful social movements that swept the industrial world in the 1960s.[32] Eventually, an alternative vision emerged from these movements, based on the respect of human dignity, the ethics of sustainability, and an ecological view of the world. This new vision forms the core of the worldwide coalition of grassroots movements.

A new kind of civil society, organized around reshaping globalization, is gradually emerging. It does not define itself vis-à-vis the state, but is global in its scope and organization. It is embodied in powerful international NGOs—such as Oxfam, Greenpeace, the Third World Network, and the Rainforest Action Network—as well as in coalitions of hundreds of smaller organizations, all of which have become social actors in a new political environment.

As political scientists Craig Warkentin and Karen Mingst point out, the new civil society is characterized by a shift of focus from formal institutions to social and political relationships among its actors.[33] These relationships are structured around two different kinds of networks. On the one hand, NGOs rely on local grassroots organizations (i.e. on living human networks); on the other hand, they skillfully use the new global communication technologies (i.e. electronic networks). The Internet, in particular, has become their most powerful political tool. By creating this unique link between human and electronic networks, the global civil society has reshaped the political landscape. To illustrate this phenomenon, Warkentin and Mingst review the recent successful anti-MAI campaign conducted by the Seattle Coalition.

The Multilateral Agreement on Investment (MAI), negotiated by the Organization for Economic Cooperation and Development (OECD), was meant to be a legal instrument that would create state-of-the-art standards for the protection of foreign investments, specifically in developing countries. Its provisions would constrain the power of governments to regulate the activities of foreign investors; by, for example, limiting restrictions on foreign ownership of real estate and even on ownership of strategic domestic industries. In other words, the sovereignty of nations would take a back seat to the rights of big business.

The negotiations began in 1995 and were conducted by the OECD behind closed doors, far from public scrutiny, for nearly two years. But in 1997, an early draft of the document was leaked to Public Citizen, a public interest group founded by Ralph Nader, which immediately published it on the Internet. As soon as this working document became publicly available (two years before Seattle), over 600 organizations in seventy countries vehemently expressed their opposition to the treaty. Oxfam, in particular, criticized the lack of transparency in the negotiation process, the exclusion of developing countries from the negotiations (even though they would be the ones most affected by the MAI), and the lack of independent reviews of the agreement's social and environmental implications.

Subsequently, the NGOs participating in the campaign posted successive drafts of the MAI on their web sites together with their own analyses, fact sheets and calls to action (including letter-writing campaigns and public demonstrations). This information appeared on numerous web sites that were extensively interlinked. Eventually, the OECD was forced to establish its own MAI web site in a largely futile effort to counter the vigorous online anti-MAI campaign.

The delegates participating in the negotiations had intended to complete the agreement in May 1997, but, in view of the well-organized worldwide opposition, the OECD declared a six-month "period of assessment" and postponed the completion date by one year. When negotiations resumed in October 1997, the chances of a successful completion had diminished drastically, and two months later the OECD announced the permanent suspension of the MAI talks. The French delegation, one of the first to withdraw its support, explicitly acknowledged the decisive role the new civil society had played in the whole process: "The MAI . . . marks [an important] step in international . . . negotiations. For the first time, we are witnessing the emergence of a 'global civil society' represented by nongovernmental organizations, which are often active in several countries and communicate across borders. This is no doubt an irreversible change."[34]

Warkentin and Mingst emphasize in their analysis that one of the principal achievements of the NGOs was to frame the public MAI discourse. Whereas the treaty was discussed in financial and economic terms by the OECD delegates, the NGOs used language that highlighted its underlying values. In doing so, they introduced a broad systemic perspective while at the same time adopting a more direct, frank and emotionally charged discourse.[35] This is typical of the new civil society, which not only uses global networks of communication but is also rooted in local communities that derive their identities from shared values.

This analysis is consistent with Manuel Castells's assertion that political power in the network society derives from the ability to use symbols and cultural codes effectively for framing the political discourse.[36] This is exactly the strength of the NGOs in the global civil society. They are able to frame critical issues in a language that makes sense to people and connects with them emotionally to promote "a more 'people-centred' politics and [more] democratic and participatory political processes."[37] As Castells concludes, the new politics "will be a cultural politics that . . . is predominantly enacted in the space of media and fights with symbols, yet connects to values and issues that spring from people's life experience."[38]

To place the political discourse within a systemic and ecological perspective, the global civil society relies on a network of scholars, research institutes, think tanks, and centers of learning that largely operate outside our leading academic institutions, business organizations and government agencies. Their common characteristic is that they pursue their research and teaching within an explicit framework of shared core values.

There are dozens of these institutions of research and learning in all parts of the world today. The best known include, in the United States, the Worldwatch Institute, the Rocky Mountain Institute, the Institute for Policy Studies, the International Forum on Globalization, Global Trade Watch, the Foundation on Economic Trends, the Institute for Food and Development Policy, the Land Institute and the Center for Ecoliteracy; Schumacher College in the U.K.; the Wuppertal Institute for Climate, Environment, and Energy in Germany; Zero Emissions Research and Initiatives in Japan, Africa, and Latin America; and the Research Foundation for Science, Technology, and Ecology in India. All these institutions have their own web sites and are interlinked with one another and with the more activist-oriented NGOs for whom they provide the necessary intellectual resources.

Most of these research institutes are communities of both scholars and activists who are engaged in a wide variety of projects and campaigns—from electoral reform to women's issues, the Kyoto Protocol on global warming, biotechnology, renewable energy, drug patents of the pharmaceutical industry etc. Among all these issues there are three clusters that seem to be focal points for the largest and most active grassroots coalitions. One is the challenge of reshaping the governing rules and institutions of globalization; the second is the opposition to genetically modified (GM) foods and the promotion of sustainable agriculture; and the third is ecodesign—a concerted effort to redesign our physical structures, cities, technologies, and industries so as to make them ecologically sustainable.

These three issue clusters are conceptually interlinked. Prohibiting the patenting of life-forms, rejecting GM foods, and promoting sustainable agriculture, for example, are important in reformulating the rules of globalization. They are essential strategies for moving toward ecological sustainability and are therefore closely linked to the broader field of ecodesign. These conceptual links mean that there are many coordinated actions among the NGOs that focus on various parts of the three issue clusters or include them in their projects.

Reshaping Globalization

Even before the Seattle teach-in in November 1999, the leading NGOs in the Seattle Coalition had formed an "Alternatives Task Force" under the leadership of the International Forum on Globalization (IFG) to synthesize the key ideas about alternatives to the current form of economic globalization. In addition to IFG, the Task Force included the Institute for Policy Studies (U.S.), Global Trade Watch (U.S.), the Council of Canadians (Canada), Focus on the Global South (Thailand and Philippines), the Third World Network (Malaysia), and the Research Foundation for Science, Technology, and Ecology (India).

After more than two years of meetings, the Task Force put together a draft interim report, "Alternatives to Economic Globalization," which was continually enriched by comments and suggestions from scholars and activists around the world, especially after the World Social Forum in Porto Alegre. The Alternatives Task Force plans to release its interim report during January 2002 and will then initiate a two-year process of refining it further through dialogues and workshops with grassroots activists around the world. The final report will be released in 2003.[39]

The IFG synthesis of alternatives to economic globalization contrasts the values and organizing principles underlying the neoliberal

Washington Consensus with a set of alternative principles and values. These include a shift from governments serving corporations to governments serving people and communities; the creation of new rules and structures that favor the local and follow the principle of subsidiarity ("Whenever power can reside at the local level, it should reside there"); the respect of cultural integrity and diversity; a strong emphasis on food security (local self-reliance in food production) and food safety (the right to healthy and safe food); as well as core labor, social, and other human rights.

The Alternatives report makes it clear that the Seattle Coalition does not oppose global trade and investment, provided that they help build healthy, respected, and sustainable communities. However, it emphasizes that the recent practices of global capitalism have shown that we need a set of rules stating explicitly that certain goods and services should not be commodified, traded, patented, or subjected to trade agreements.

In addition to already existing rules of this kind, which concern endangered species and goods that are harmful to the environment or to public health and safety—toxic waste, nuclear technology, armaments, etc.—the new rules would also concern goods that belong to the global commons, that is, goods that are part of the fundamental building blocks of life or of humanity's common inheritance. Included in this are goods like bulk fresh water, which should not be traded but should be given away to those in need; seeds, plants, and animals that are traded in traditional farming communities but should not be patented for profit; and DNA sequences that should neither be patented nor traded.

The authors of the report acknowledge that these issues constitute perhaps the most difficult, but also the most important, part of the globalization debate. Their main concern is to stem the tide of a global trading system where everything is for sale, even our biological heritage, or access to seeds, food, air, and water—elements of life that were once considered sacred.

In addition to the discussions of alternative values and organizing principles, the IFG synthesis includes concrete, and radical, proposals for restructuring the Bretton Woods institutions. Most of the NGOs in the Seattle Coalition feel that reforming the WTO, the World Bank, and the IMF is not a viable strategy, because their structures, mandates, purposes, and operating processes are fundamentally at odds with the core values of human dignity and ecological sustainability. Instead, the NGOs propose a four-part restructuring process: dismantling the Bretton Woods institutions, unifying global governance under a reformed United Nations system, strengthening certain existing UN organizations, and creating several new organizations within the UN that would fill the gap left by the Bretton Woods institutions.

The report points out that we now have two strikingly different sets of institutions of global governance: the Bretton Woods triad and the United Nations. The Bretton Woods institutions have been more effective in implementing well-defined agendas, but these have been largely destructive and have been imposed on humanity in coercive, undemocratic ways. The United Nations, by contrast, has been less effective but its mandate is much broader; its decision-making processes are more open and democratic; and its agendas give much greater weight to social and environmental priorities. The NGOs argue that limiting the powers and mandates of the IMF, World Bank, and WTO will create space for a reformed United Nations to fulfill its intended functions.

The Seattle Coalition proposes that any plans for a new round of WTO negotiations or for any expansion of the WTO mandate or membership should be firmly rejected. Instead, the power of the WTO should be either eliminated or radically reduced to make it simply one among many international organizations in a pluralistic world with multiple checks and balances. As the campaign launched by Global Trade Watch puts it, "WTO: Shrink it or Sink it."

As for the World Bank and the IMF, the Seattle Coalition believes that these institutions bear major responsibility for burdening Third World countries with unpayable foreign debts and for implementing a misguided concept of development that has had disastrous social and ecological consequences. Borrowing a phrase applied to ageing nuclear power plants, the report suggests that it is time to "decommission" the Bank and the IMF.

To carry out the original mandates of the Bretton Woods institutions, the Alternatives report proposes to strengthen the mandates and resources of existing UN organizations like the World Health Organization, the International Labor Organization, and the UN Environment Program. Its authors believe that instead of placing trade-related health, labor, and environmental standards under the jurisdiction of the WTO, they should be placed under the authorities of UN agencies and given priority over trade expansion. In the view of the Seattle Coalition, public health, workers' rights, and environmental protection are ends in themselves, whereas international trade and investment are only means.

In addition, the Alternatives report supports the creation of a small number of new global institutions under UN authority and oversight. These include an International Insolvency Court (IIC) to oversee debt relief, which would become operative as the World Bank and regional development banks are decommissioned; an International Finance Organization (IFO), which would replace the IMF and would work with UN member countries to achieve and maintain balance and stability in international financial relationships; and an Organization for Corporate Accountability (OCA) under the mandate and direction of the United Nations. The primary function of the OCA would be to provide governments and the general public with comprehensive and authoritative information about corporate practices in support of negotiations of relevant bilateral and multilateral agreements, as well as investor and consumer boycotts.

The main thrust of all these proposals is to decentralize the power of global institutions in favor of a pluralistic system of regional and international organizations, each of which would be checked by other organizations, agreements, and regional groupings. It seems that such a less structured and more fluid system of global governance is much more appropriate for today's world, in which corporations are increasingly organized as decentralized networks and political authority is shifting to regional and local levels as nation-states transform themselves into network states.[40]

In conclusion, the Alternatives report points out that its proposals would have seemed quite unrealistic a few years ago, but that the political landscape has changed dramatically since Seattle. The Bretton Woods institutions are mired in a deep crisis of legitimacy, and an alliance of southern countries (the "G-77 nations"), sympathetic politicians from the North and the new global civil society may well emerge with sufficient power to achieve sweeping institutional reforms and reshape globalization.

The Food Revolution

Unlike the protests against economic globalization, the resistance against genetically modified foods did not begin with a campaign of public education. It began in the early 1990s with widespread demonstrations by traditional farmers in India, followed by consumer boycotts in Europe, combined with a spectacular renaissance of organic farming. In the words of environmental health activist and author John Robbins: "All over the world, people were calling for their governments to protect human welfare and the environment, rather than put corporate profits over public health. People everywhere were insisting on a society that restores the Earth, not one that destroys it."[41]

Boycotts and demonstrations directed against various biotech and agrochemical corporations were soon followed by extensive documentation of the industry's practices by the leading NGOs in the ecology and environmental health movements.[42]

In his richly documented book *The Food Revolution*, John Robbins gives a vivid account of the citizen revolt against GM foods that rapidly spread from Europe to the rest of the world.[43] In 1998, genetically engineered crops were destroyed by angry citizens and farmers in Great Britain, Ireland, France, Germany, the Netherlands, and Greece, as well as in the United States, India, Brazil, Australia, and New Zealand. At the same time, grassroots groups around the world organized massive petitions to their governments. In Austria, for example, over a million citizens, representing 20 percent of the electorate, signed a petition to ban GM foods. In the United States, a petition to demand mandatory labeling of transgenic food was signed by half a million people and presented to Congress; and throughout the world, countless organizations, including the British Medical Association, called for a moratorium on all crops containing genetically modified organisms (GMOs).

Governments soon responded to these forceful expressions of public opinion. The governor of Brazil's major soybean-growing state, Rio Grande do Sul, which hosted the World Social Forum in Porto Alegre, declared the entire state a GMO-free zone. The governments of France, Italy, Greece, and Denmark announced that they would block the approval of new GM crops in the European Union. The European Commission made the labeling of GM foods mandatory, as did the governments of Japan, South Korea, Australia, and Mexico. In January 2000, 130 nations signed the groundbreaking Cartagena Protocol on Biosafety in Montreal, which gives nations the right to refuse entry to any genetically modified forms of life, despite vehement opposition from the United States.

The response of the corporate community to the massive civic uprising against food biotechnology was no less decisive. Food producers, restaurants, and beverage companies all over the world were quick to pledge that they would eliminate GMOs from their products. In 1999, the seven largest grocery chains in six European countries made a public commitment to go "GMO-free," and were followed in this commitment within days by the huge food companies Unilever (which had been one of the most aggressive proponents of GM foods), Nestlé, and Cadbury-Schweppes.

At the same time, Japan's two largest breweries, Kirin and Sapporo, announced that they would not use genetically modified corn in their beer. Subsequently, the fast-food chains McDonald's and Burger King told their suppliers that they would not buy any more genetically altered potatoes. GM potatoes were also phased out by major manufacturers of potato chips, while Frito-Lay told its corn farmers to stop supplying GM corn.

As the food industry increasingly turned away from GM foods and the acreage of transgenic crops began to shrink, reversing the explosive growth of the late nineties, analysts naturally began to warn investors about the financial risks of food biotechnology. In 1999, Europe's largest bank, Deutsche Bank, declared categorically that "GMOs are dead" and recommended that its clients sell all their holdings in biotech companies.[44] One year later, the *Wall Street Journal* came to the same conclusion: "With the controversy over genetically modified foods spreading across the globe and taking a toll on the stocks of companies with agricultural-biotechnology business, it's hard to see those companies as a good investment, even in the long run."[45] These recent developments show clearly that today's worldwide grassroots movements have the power and the skills to change not only the international political climate, but also the game of the global market, by reorienting its financial flows according to different values.

Ecoliteracy and Ecodesign

Ecological sustainability is an essential component of the core values that form the basis for reshaping globalization. Accordingly, many of the NGOs, research institutes, and centers of learning in the new global civil society have chosen sustainability as their explicit focus. Indeed, creating sustainable communities is the great challenge of our time.

The concept of sustainability was introduced in the early 1980s by Lester Brown, founder of the Worldwatch Institute, who defined a sustainable society as one that is able to satisfy its needs without diminishing the chances of future generations.[46] Several years later, the report of the World Commission on Environment and Development (the "Brundtland Report") used the same definition to present the notion of sustainable development: "Humankind has the ability to achieve sustainable development—to meet the needs of the present without compromising the ability of future generations to meet their own needs."[47] These definitions of sustainability are important moral exhortations. They remind us of our responsibility to pass on to our children and grandchildren a world with as many opportunities as the one we inherited. However, this definition does not tell us anything about how to build a sustainable society. This is why there has been much confusion about the meaning of sustainability, even within the environmental movement.

The key to an operational definition of ecological sustainability is the realization that we do not need to invent sustainable human communities from scratch but can model them after nature's ecosystems, which are sustainable communities of plants, animals, and microorganisms. Since the outstanding characteristic of the Earth household is its inherent ability to sustain life,[48] a sustainable human community is one designed in such a manner that its ways of life, businesses, economy, physical structures, and technologies do not interfere with nature's inherent ability to sustain life. Sustainable communities evolve their patterns of living over time in continual interaction with other living systems, both human and nonhuman. Sustainability does not mean that things do not change: it is a dynamic process of coevolution rather than a static state.

The operational definition of sustainability implies that the first step in our endeavor to build sustainable communities must be to become "ecologically literate," i.e., to understand the principles of organization, common to all living systems, that ecosystems have evolved to sustain the web of life.[49] As we have seen throughout this book, living systems are self-generating networks, organizationally closed within boundaries but open to continual flows of energy and matter. This systemic understanding of life allows us to formulate a set of principles of organization that may be identified as the basic principles of ecology and used as guidelines for building sustainable human communities. Specifically, there are six principles of ecology that are critical to sustaining life: networks, cycles, solar energy, partnership, diversity and dynamic balance (see table opposite).

These principles are directly relevant to our health and well-being. Because of our vital need to breathe, eat, and drink, we are always embedded in the cyclical processes of nature. Our health depends upon the purity of the air we breathe and the water we drink, and it depends on the health of the soil from which our food is produced. In the coming decades the survival of humanity will depend on our ecological lit-

PRINCIPLES OF ECOLOGY

Networks
At all scales of nature, we find living systems nesting within other living systems—networks within networks. Their boundaries are not boundaries of separation but boundaries of identity. All living systems communicate with one another and share resources across their boundaries.

Cycles
All living organisms must feed on continual flows of matter and energy from their environment to stay alive, and all living organisms continually produce waste. However, an ecosystem generates no net waste, one species' waste being another species' food. Thus, matter cycles continually through the web of life.

Solar Energy
Solar energy, transformed into chemical energy by the photosynthesis of green plants, drives the ecological cycles.

Partnership
The exchanges of energy and resources in an ecosystem are sustained by pervasive cooperation. Life did not take over the planet by combat but by cooperation, partnership, and networking.

Diversity
Ecosystems achieve stability and resilience through the richness and complexity of their ecological webs. The greater their biodiversity, the more resilient they will be.

Dynamic Balance
An ecosystem is a flexible, ever-fluctuating network. Its flexibility is a consequence of multiple feedback loops that keep the system in a state of dynamic balance. No single variable is maximized; all variables fluctuate around their optimal values.

eracy—our ability to understand the basic principles of ecology and to live accordingly. Thus, ecological literacy, or "ecoliteracy," must become a critical skill for politicians, business leaders, and professionals in all spheres, and should be the most important part of education at all levels—from primary and secondary schools to colleges, universities, and the continuing education and training of professionals.

At the Center for Ecoliteracy in Berkeley (www.ecoliteracy.org), my colleagues and I are developing a system of education for sustainable living, based on ecological literacy, at the primary and secondary school levels.[50] This involves a pedagogy that puts the understanding of life at its very center; an experience of learning in the real world (growing food, exploring a watershed, restoring a wetland) that overcomes our alienation from nature and rekindles a sense of place; and a curriculum that teaches our children the fundamental facts of life—that one species' waste is another species' food; that matter cycles continually through the web of life; that the energy driving the ecological cycles flows from the sun; that diversity assures resilience; that life, from its beginning more than 3 billion years ago, did not take over the planet by combat but by networking.

This new knowledge, which is also ancient wisdom, is now being taught within a growing network of schools in California, and is beginning to spread to other parts of the world. Similar efforts are under way in higher education, pioneered by Second Nature (www.secondnature.org), an educational organization in Boston that collaborates with numerous colleges and universities to make education for sustainability an integral part of campus life.

In addition, ecological literacy is being transmitted and continually refined in informal teach-ins and in the new institutions of learning of the emerging global civil society. Schumacher College, in England, is an outstanding example of such new institutions. It is a center for ecological studies with philosophical and spiritual roots in deep ecology, where students from all parts of the world gather to learn, live, and work together while being taught by an international faculty.

Ecoliteracy—the understanding of the principles of organization that ecosystems have evolved to sustain the web of life—is the first step on the road to sustainability. The second step is to move toward ecodesign. We need to apply our ecological knowledge to the fundamental redesign of our technologies and social institutions, so as to bridge the current gap between human design and the ecologically sustainable systems of nature.

Fortunately, this is already taking place. In recent years, there has been a dramatic rise in ecologically oriented design practices and projects. The recently published book *Natural Capitalism*, by Paul Hawken and Amory and Hunter Lovins, provides comprehensive overall documentation, and the Lovinses' Rocky Mountain Institute (www.rmi.org) serves as a clearinghouse for up-to-date information on a wide variety of ecodesign projects.

Design, in the broadest sense, consists in shaping flows of energy and materials for human purposes. Ecodesign is a process in which our human purposes are carefully meshed with the larger patterns and flows of the natural world. Ecodesign principles reflect the principles of organization that nature has evolved to sustain the web of life. To practice industrial design in such a context requires a fundamental shift in our attitude toward nature. In the words of science writer Janine Benyus, it "introduces an era based not on what we can *extract* from nature, but on what we can *learn* from her."[51]

When we speak of the "wisdom of nature," or of the marvelous "design" of a butterfly's wings, or a spider's silk thread, we need to remember that our language is metaphorical.[52] However, this does not change the fact that, from the perspective of sustainability, nature's "design" and "technologies" are far superior to human science and technology. They were created and have been continually refined over billions of years of evolution, during which the inhabitants of the Earth household flourished and diversified without ever using up their natural capital—the planet's resources and ecosystem services on which the well-being of all living creatures depends.

Ecological Clustering of Industries

The first principle of ecodesign is that "waste equals food." Today, a major clash between economics and ecology derives from the fact that nature's ecosystems are cyclical, whereas our industrial systems are linear. In nature, matter cycles continually, and thus ecosystems generate no overall waste. Human businesses, by contrast, take natural resources, transform them into products plus waste, and sell the products to consumers, who discard more waste when they have used the products.

The principle "waste equals food" means that all products and materials manufactured by industry, as well as the wastes generated in the manufacturing processes, must eventually provide nourishment for something new.[53] A sustainable business organization would be embedded in an "ecology of organizations," in which the waste of any one organization would be a resource for another. In such a sustainable industrial system, the total outflow of each organization—its products *and* wastes—would be perceived and treated as resources cycling through the system.

Such ecological clusters of industries have actually been initiated in many parts of the world by an organization called Zero Emissions Research and Initiatives (ZERI), founded by business entrepreneur Gunter Pauli in the early 1990s. Pauli introduced the notion of industrial clustering by promoting the principle of zero emissions and making it the very core of the ZERI concept. Zero emissions means zero waste. Taking nature as its model and mentor, ZERI strives to eliminate the very idea of waste.

To appreciate how radical an approach this is, we need to realize that our current businesses throw away most of the resources they take from nature. For example, when we extract cellulose from wood to make paper, we cut down forests but use only 20 to 25 percent of the trees, discarding the remaining 75 to 80 percent as waste. Beer breweries extract only 8 percent of the nutrients from barley or rice for fermentation; palm oil is a mere 4 percent of the palm tree's biomass; and coffee beans are 3.7 percent of the coffee bush.[54]

Pauli's starting point was to recognize that the organic waste that is thrown away or burned by one industry contains an abundance of precious resources for other industries. ZERI helps industries to organize themselves into ecological clusters, so that the waste of one can be sold as a resource to another, for the benefit of both.[55]

The principle of zero emissions ultimately implies zero material consumption. Like nature's ecosystems, a sustainable human community would use energy that flows from the sun but would not consume any material goods without recycling them after use. In other words, it would not use any new materials. Moreover, zero emissions also means no pollution. ZERI's ecological clusters are designed to operate in an environment free of toxic wastes and pollution. Thus "waste equals food," the first principle of ecodesign, points to the ultimate solution for some of our major environmental problems.

From the economic point of view, the ZERI concept means a vast increase in resource productivity. According to classical economic theory, productivity results from the effective combination of three sources of wealth: natural resources, capital, and labor. In the current economy, business leaders and economists concentrate mainly on capital and labor to increase productivity, creating economies of scale with disastrous social and environmental consequences.[56] The ZERI concept implies a shift from labor productivity to resource productivity, since waste is transformed into new resources. Ecological clustering dramatically increases productivity and improves product quality, while at the same time creating jobs and reducing pollution.

The ZERI organization is an international network of scholars, business people, government officials, and educators.[57] The scholars play a key role, because the organization of the industrial clusters is based on detailed knowledge of the biodiversity and biological processes in local ecosystems. Pauli originally launched ZERI as a research project at the United Nations University in Tokyo. To do so, he created a network of scientists on the Internet, using the existing academic networks of the Royal Swedish Academy of Sciences, the Chinese Academy of Sciences, and the Third World Academy of Sciences. Being one of the first to pioneer scientific exchanges and conferences on the Internet, Pauli excited the scientists' interest, and by continually asking them challenging questions about biochemistry, ecology, climatology, and other disciplines, he generated not only business solutions but also numerous new ideas for scientific research. To emphasize the Socratic nature of this method, he called ZERI's first academic network Socrates Online. Since then, the ZERI network of researchers has grown to 3,000 scholars worldwide.

ZERI has now initiated some fifty projects around the world and operates twenty-five project centers on five continents in very diverse climates and cultural settings. The clusters around Colombian coffee farms are good illustrations of the basic ZERI method. These farms are in crisis because of the dramatic drop in the price of coffee beans on the world market. Meanwhile, the farmers use only 3.7 percent of the coffee plant, returning most of the waste to the environment as landfill and pollution—smoke, waste water, and caffeine-contaminated compost. ZERI put this waste to work. Research showed that coffee biomass can be used profitably to cultivate tropical mushrooms, feed livestock, compost organic fertilizer, and generate energy. The resulting ZERI cluster is pictured opposite.

The waste of each component in the cluster provides a resource for another component. To put it in greatly simplified terms, when the coffee beans are harvested, the remains of the coffee plant are used to grow shiitake mushrooms (a high-priced delicacy); the remains of the mushrooms (rich in protein) feed earthworms, cattle, and pigs; earthworms feed chickens; cattle and pig manure produces biogas and sludge; the sludge fertilizes the coffee farm and surrounding vegetable gardens, while the energy from the biogas is used in the process of mushroom farming.

The clustering of these productive systems inexpensively generates several revenue streams in addition to the original coffee beans—from poultry, mushrooms, vegetables, beef, and pork—while creating jobs in the local community. The results are beneficial both to the environment and the community; there are no high investments; and there is no need for the coffee farmers to give up their traditional livelihood.

Technologies in the typical ZERI clusters are small-scale and local.

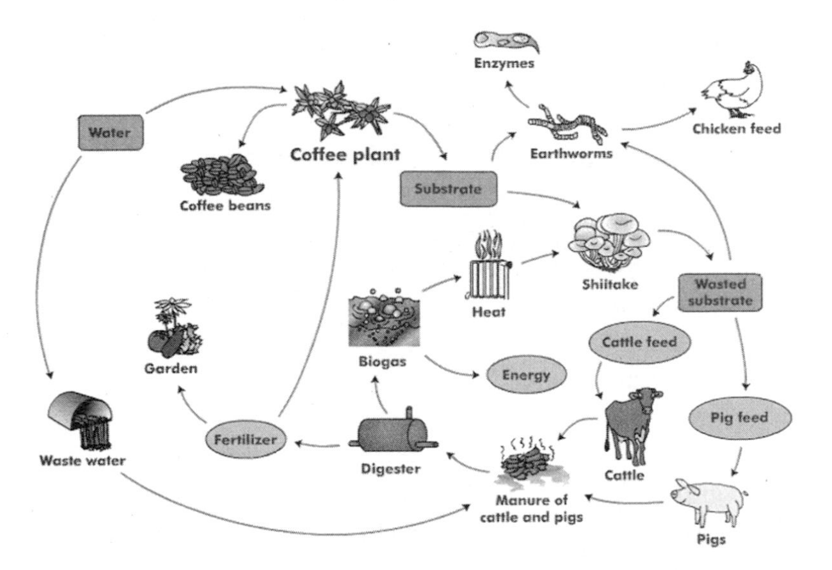

Ecological cluster around a Colombian coffee farm (from www.zeri.org)

The places of production are usually close to those of consumption, which eliminates or radically reduces transportation costs. No single production unit tries to maximize its output, because this would only unbalance the system. Instead, the goal is to optimize the production processes of each component, while maximizing the productivity and ecological sustainability of the whole.

Similar agricultural clusters, with beer breweries as their center instead of coffee farms, are operating in Africa, Europe, Japan, and other parts of the world. Other clusters have aquatic components; for example, a cluster in southern Brazil includes the farming of highly nutritious spirulina algae in the irrigation channels of rice fields (which otherwise are used only once a year). The spirulina is used as special enrichment in a "ginger cookie" program in rural schools to fight widespread malnutrition. This generates additional revenue for the rice farmers while responding to a pressing social need.

An impressive realization of the ZERI concept on a large scale is the reforestation project of the environmental research center Las Gaviotas in eastern Colombia, established and directed by ecodesigner

Paolo Lugari. In the midst of Colombia's deep social crisis, Las Gaviotas has created an environment full of innovation and hope.

When ZERI arrived at Las Gaviotas, the center had already established a worldwide reputation through the development of many ingenious renewable-energy technologies, including solar water heating for thousands of housing units in the capital, Bogotá, as well as a rural hospital that produces its own solar energy, distills its own water and cooks locally grown food.

After these successes, Lugari embarked on the most extensive reforestation program Colombia has ever seen. Growing trees in the eastern savannas (the *lanos*) is a massive challenge. High soil acidity and extreme temperatures severely limit the choices of young trees that might survive the hot, dry summers. However, after careful analysis, the scientists at Las Gaviotas concluded that a species known as Caribbean pine would be able to adapt to these extreme conditions.

After the first two years of planting they proved to be correct, and since then, the center has planted thousands of hectares with the help of specially developed tree-planting machines. At first, there was concern that such a vast monoculture of pine trees might have adverse ecological consequences, but the opposite occurred. As pine needles dropped continually on the forest floor, they created a rich cover of humus, which made it possible for new plants, trees, and forest undergrowth to thrive. Today, over 200 new species are found in this microclimate that do not grow anywhere else in the savannah. And with these new plant species come bacteria, insects, birds, and even mammals. Biodiversity has increased dramatically.

In addition to drawing CO_2 from the air (which helps reduce global warming) and recovering lost biodiversity, the pine forest also produces lucrative colofonia sap, which is collected and processed into a prime ingredient for the production of natural paints and high-quality glossy paper. This creates further employment and valuable revenue streams. Finally, it turned out that the bacteria generated in the newly planted forest act as an excellent filtering system, purifying the subsoil water, which also happens to be rich in minerals. The center collects and bottles the resulting mineral water at very low cost. This provides an important means for preventive health care, since most of the region's health problems stem from poor water quality. The success story of Las Gaviotas is a powerful demonstration of the ZERI concept. Driven by the reforestation program, the ecological cluster—designed collaboratively by a ZERI/Las Gaviotas team—has helped reduce global warming, increased biodiversity, created jobs for the local indigenous population, generated new revenue streams, and contributed significantly to the improvement of public health in the region.

In building up the ZERI organization, Gunter Pauli used the most advanced techniques of electronic networking and conferencing. ZERI consists of three types of interconnected networks. One is the ecological cluster of industries, patterned after the food webs in nature's ecosystems. Closely associated is the human network of the local community where the cluster is located. The third, finally, is the international network of scientists who provide the detailed knowledge necessary to design clusters of industries that are compatible with local ecosystems, climatic conditions and cultural settings. Due to the nonlinear nature of these interconnected networks, the solutions they produce are multiple, or "systemic," solutions. The combined value created by the whole is always greater than the sum of the values that would be generated by independently operating components.

Because of their sharp increases in resource productivity, these clustered industries can aim for quality levels in their products that are considerably higher than those that corresponding stand-alone businesses can afford. As a consequence, the ZERI businesses are competitive on the global market—not in the sense that they sell their products globally, but in that nobody can compete with them on their home turf. As in ecosystems, diversity increases resilience. The more diverse the ZERI clusters become, the more resilient and competitive they are. Theirs is not an economy of scale but, as Pauli puts it, an "economy of scope."

It is not difficult to see that the principles of organization underlying the ZERI concept—the nonlinear network structure, cycling of matter, multiple partnerships, diversity of enterprises, local production and consumption, and the goal of optimizing instead of maximizing—are basic principles of ecology. This is, of course, not coincidental. The ZERI clusters are impressive examples of ecoliteracy embodied in ecodesign.

An Economy of Service and Flow

Most of the ZERI clusters involve organic resources and wastes. To build sustainable industrial societies, however, the ecodesign principle "waste equals food" and the resulting cycling of matter must extend beyond organic products. This concept has been best articulated by ecodesigners Michael Braungart in Germany and William McDonough in the United States.[58]

Braungart and McDonough speak of two kinds of metabolisms—a biological metabolism and a technical metabolism. Matter that cycles in the biological metabolism is biodegradable and becomes food for other living organisms. Materials that are not biodegradable are regarded as technical nutrients, which continually circulate within industrial cycles that constitute the technical metabolism. In order for these two metabolisms to remain healthy, great care must be taken to keep them distinct and separate, so that they do not contaminate each other. Things that are part of the biological metabolism—agricultural products, clothing, cosmetics, etc.—should not contain persistent toxic substances. Things that go into the technical metabolism—machines, physical structures, etc.—should be kept well apart from the biological metabolism.

In a sustainable industrial society, all products, materials, and wastes will either be biological or technical nutrients. Biological nutrients will be designed to reenter ecological cycles to be consumed by microorganisms and other creatures in the soil. In addition to organic waste from our food, most packaging (which makes up about half the volume of our solid-waste stream) should be composed of biological nutrients. With today's technologies, it is quite feasible to produce packaging that can be tossed into the compost bin to biodegrade. As McDonough and Braungart point out, "There is no need for shampoo bottles, toothpaste tubes, yogurt cartons, juice containers, and other packaging to last decades (or even centuries) longer than what came inside them."[59]

Technical nutrients will be designed to go back into technical cycles. Braungart and McDonough emphasize that the reuse of technical nutrients in industrial cycles is distinct from conventional recycling, because it maintains the high quality of the materials, rather than "downcycling" them into flower pots or park benches. Technical metabolisms equivalent to the ZERI clusters have not yet been established, but there is definitely a trend to do so. In the United States, which is not a world leader in recycling, more than half of its steel is now produced from scrap. Similarly, there are more than a dozen paper mills running only on waste paper in the state of New Jersey alone.[60] The new steel mini-mills do not need to be located near mines, nor the paper mills near forests. They are located near the cities that produce the waste and consume the raw materials, which saves considerable transportation costs.

Many other ecodesign technologies for the repeated use of technical nutrients are on the horizon. For example, it is now possible to create special types of ink that can be removed from paper in a hot water bath without damaging the paper fibers. This chemical innovation would allow complete separation of paper and ink so that both can be reused. The paper would last ten to thirteen times longer than conventionally recycled paper fibers. If this technique were universally adopted, it could reduce the use of forest pulp by 90 percent, in addition to reducing the amounts of toxic ink residues that now end up in landfills.[61]

If the concept of technical cycles were fully implemented, it would lead to a fundamental restructuring of economic relationships. After all, what we want from a technical product is not a sense of ownership but the service the product provides. We want entertainment from our VCR, mobility from our car, cold drinks from our refrigerator, and so on. As Paul Hawken likes to point out, we do not buy a television set in order to own a box of 4,000 toxic chemicals; we do so in order to watch television.[62]

From the perspective of ecodesign, it makes no sense to own these products and to throw them away at the end of their useful lives. It makes much more sense to buy their services, i.e. to lease or rent them. Ownership would be retained by the manufacturer, and when one had finished using a product, or wanted to upgrade to a newer version, the manufacturer would take the old product back, break it down into its basic components—the technical nutrients—and use those in the assembly of new products, or sell them to other businesses.[63] The resulting economy would no longer be based on the ownership of goods but would be an economy of service and flow. Industrial raw materials and technical components would continually cycle between manufacturers and users, as they would between different industries.

This shift from a product-oriented economy to a service-and-flow economy is no longer pure theory. One of the world's largest carpet manufacturers, a company called Interface, based in Atlanta, has begun the transition from selling carpets to leasing carpeting services.[64] The basic idea is that people want to walk on and look at a carpet, not own it. They can obtain those services at much lower cost if the company owns the carpet and remains responsible for keeping it in good shape in exchange for a monthly fee. Interface carpets are laid in the form of tiles, and only tiles that are worn are replaced after a regular monthly inspection. This reduces not only the amount of carpet material needed for replacements, but also minimizes disruptions, because the worn tiles are usually not found under furniture. When a customer wants to replace the entire carpet, the company takes it back, extracts its technical nutrients, and provides the customer with a new carpet in the desired color, style, and texture.

These practices, together with several innovations in materials design, have made Interface one of the pioneers of the new service-and-flow economy. Similar innovations have been undertaken in the photocopying industry by Canon, in Japan, and in the automotive industry by Fiat, in Italy. Canon has revolutionized the photocopying industry by redesigning its copiers so that more than 90 percent of their components can be reused or recycled.[65] In Fiat's Auto Recycling (FARE) system, the steel, plastics, glass, seat padding, and many other components of old Fiat cars are retrieved in over 300 dismantling centers, to be reused in new cars or passed on as resources to other industries. The company has established a target of 85 percent recycling of materials by 2002 and of 95 percent by 2010. The Fiat program has also been extended from Italy to other European countries and to Latin America.[66]

In a service-and-flow economy manufacturers must be able to take their products apart easily in order to redistribute the raw materials. This will have a major impact on product design. The most successful products will be those which contain a small number of materials and components that can easily be disassembled, separated, rearranged, and reused. The companies mentioned above have all completely redesigned their products for easy dismantling. When this happens, the demand for labor (to do all the disassembling, sorting, and recycling) will increase, as waste decreases. Thus, the service-and-flow economy involves a shift from natural resources, which are scarce, to human resources, which are abundant.

Another effect of this new product design will be to align the interests of manufacturers and customers when it comes to product durability. In an economy based on selling goods, the obsolescence and frequent disposal and replacement of those goods is in the manufacturers' financial interests, even though that is harmful to the environment and costly for the customers. In a service-and-flow economy, by contrast, it is in the interest of both manufacturers and customers to create long-living products while using a minimum of energy and materials.

Doing More with Less

Even though the complete cycling of materials in technical clusters has not yet been achieved, existing partial clusters and material loops have led to dramatic increases in energy and resource efficiency. Ecodesigners today are confident that an astonishing *90 percent reduction* in energy and materials—called Factor Ten because it corresponds to a tenfold increase in resource efficiency—is possible in developed countries with existing technologies and without any decline in people's living standards.[67] The environment ministers of several European countries, as well as the United Nations Environment Program (UNEP), have urged adoption of Factor Ten goals.[68]

Such dramatic increases in resource productivity are made possible by the massive inefficiency and waste that are characteristic of most current industrial design. As in the case of biological resources, ecodesign principles such as networking, recycling, and optimizing instead of maximizing have not been part of the theory and practice of industrial design, and resource productivity has not even been part of designers' vocabulary until very recently.

Natural Capitalism, by Paul Hawken, Amory Lovins, and Hunter Lovins, is full of astounding examples of dramatic increases in resource efficiency. The authors estimate that by pursuing these efficiencies we could almost halt the degradation of the biosphere, and emphasize that the present massive inefficiencies almost always cost more than the measures that would reverse them.[69] In other words, ecodesign is good business. As in the ZERI clusters, the increase of resource productivity in the technical sphere has multiple beneficial effects. It slows the depletion of natural resources, reduces pollution, and increases employment. Resource productivity alone will not solve our environmental crisis, but it can buy us precious time to allow for the transition to a sustainable society.

One area where ecodesign has led to a wide range of impressive innovations is in the design of buildings.[70] A well-designed commercial structure will display a physical shape and orientation that takes the greatest advantage of the sun and wind, optimizing passive solar heating and cooling. That alone will usually save about one third of the building's energy use. Proper orientation, combined with other passive solar design features, also provides glare-free natural light throughout the structure, which usually provides sufficient lighting during daytime. Modern electric lighting systems can produce pleasant and accurate colors and eliminate all flicker, hum, and glare. Typical energy savings from such lighting are between 80 and 90 percent, which usually pays for the installation of the lighting systems within one year.

Perhaps even more impressive are the dramatic improvements in insulation and temperature regulation created by "superwindows," which keep people warm in winter and cool in summer without any additional heating or cooling. Superwindows are covered with several invisible coatings that let through light but reflect heat, in addition to having double panes, the space between which is filled with heavy gas that blocks the flow of heat and noise. Experimental buildings equipped with superwindows have shown that complete comfort can be maintained without any heating or cooling equipment, even with outdoor conditions ranging from severe cold to extreme heat.

Finally, ecodesigned buildings not only save energy by letting in natural light and keeping out the weather; they can even produce energy. Photovoltaic electricity can now be generated from wall panels, roofing shingles, and other structural elements that look and work like ordinary building materials but produce electricity whenever there is sunlight, even if it comes through clouds. A building with such photovoltaic materials as roofs and windows can produce more daytime electricity than it uses. Indeed, that is what half a million solar-powered homes around the world do every day.

These are just some of the most important recent innovations in the ecodesign of buildings. They are not confined to new buildings; they can also be implemented by retrofitting old structures. The savings in energy and materials created by these design innovations are dramatic, and the buildings are also more comfortable and healthier to live and work in. As ecodesign innovations continue to accumulate, buildings will come ever closer to the vision advanced by William McDonough and Michael Braungart: "Imagine . . . a building as a kind of tree. It would purify air, accrue solar income, produce more energy than it consumes, create shade and habitat, enrich soil, and change with the seasons."[71] Several examples of buildings with some of these revolutionary features already exist today.[72]

Another sector where huge savings in energy are possible is transportation. As we have seen, the WTO's free-trade rules are designed to stifle local production in favor of exports and imports, which massively increases long-distance transportation and puts enormous stress on the environment.[73] Reversing that trend, which is an important part of the Seattle Coalition's program of reshaping globalization, will lead to massive energy savings. This can be seen already in several of the pioneering examples of ecodesign mentioned in the preceding pages—from local and small-scale ecological clusters of industries to the new mini-mills for local production of steel and paper from scrap and waste, and food from organic farms which is locally produced and sold.

Similar considerations apply to urban design. The urban and suburban sprawl that characterizes most modern cities, especially in North America, has created very high automobile dependence with a minimal role for public transport, cycling, or walking. The consequences: high consumption of gasoline and correspondingly high levels of smog, severe stress due to traffic congestion, and loss of street life, community, and public safety.

The past three decades have seen the emergence of an international "ecocity" movement, which tries to counteract urban sprawl by using ecodesign principles, to redesign our cities so that they become ecologically healthy.[74] By carefully analyzing transport and land-use patterns, urban planners Peter Newman and Jeff Kenworthy found that energy use depends critically on city density.[75] As the city becomes denser, the use of public transport and the amount of walking and cycling increase, while the use of cars decreases. Historic city centers with high density and mixed land use, which have been reconverted into the car-free environments they were originally meant to be, now exist in most European cities. Other cities have created modern car-free environments that encourage walking and cycling. These newly designed neighborhoods, known as "urban villages," display high-density structures combined with ample common green spaces.

The German city of Freiburg, for example, has an urban village called Seepark, built around a large lake and a light-rail line. The community is totally car free; all movement is on foot and bicycle; there is plenty of open space where children are safe. Similar urban villages, car free and integrated with public transportation, have been created in several other cities, including Munich, Zurich, and Vancouver. The application of ecodesign principles has brought these areas multiple benefits—significant energy savings and a healthy and safe environment with drastically reduced levels of pollution.

In addition to the developments described above, major savings in energy and materials are also being achieved through a radical redesign of automobiles, but even though "hypercars"—ultralight, superefficient, and pollution-free automobiles—will soon be on the market,[76] this will not solve the multiple health, social, and environmental problems caused by the excessive use of cars. Only fundamental changes in our patterns of production and consumption and in the design of our cities will accomplish that. In the meantime, however, hypercars, like other sharp increases in resource productivity, will significantly reduce pollution and buy us much needed time for the transition to a sustainable future.

Energy from the Sun

Before turning to the ecodesign of automobiles, we need to examine more fully the question of energy use. In a sustainable society, all human activities and industrial processes must ultimately be fueled by solar energy, like the processes in nature's ecosystems. Solar energy is the only kind of energy that is renewable and environmentally benign. Hence, the shift to a sustainable society centrally includes a shift from fossil fuels—the principal energy sources of the Industrial Age—to solar power.

The sun has supplied the planet with energy for billions of years, and virtually all our energy sources—wood, coal, petroleum, natural gas, wind, hydropower, and so on—originate in solar energy. However, not all these forms of energy are renewable. In the current energy debate, the term "solar energy" is used to refer to the forms of energy that come from inexhaustible or renewable sources—sunlight for solar heating and photovoltaic electricity, wind and hydropower and biomass (organic matter). The most efficient solar technologies involve small-scale devices, used by local communities, which generate a wide variety of jobs. The use of solar energy, like the other ecodesign principles, reduces pollution while at the same time increasing employment. Moreover, the shift to solar energy will especially benefit people living in Southern countries where sunlight is most abundant.

In recent years, it has become increasingly clear that the transition to solar energy is needed not only because fossil fuels—coal, oil and natural gas—are limited and nonrenewable, but especially because of their devastating impacts on the environment. The discovery of the critical role of carbon dioxide (CO_2) in global climate change and of humanity's responsibility for adding CO_2 to the atmosphere has highlighted the connection between environmental pollution and the carbon content of fossil-fuel energy, and carbon intensity has become an important indicator of our movement toward sustainability. As Seth Dunn of the Worldwatch Institute puts it, we need to "decarbonize" our energy economy.[77]

Fortunately, this is already happening. Industrial ecologist Jesse Ausubel, cited by Dunn, has shown that a progressive decarbonization of energy sources has taken place over the past 200 years. For millennia, humanity's main energy source was wood, which releases ten molecules of carbon (in soot or CO_2) for every molecule of hydrogen (in water vapor) when it is burned. When coal became the principal source of energy for the industrial world in the nineteenth century, that ratio was reduced to 2:1. By the mid-twentieth century, oil surpassed coal as the leading fuel. This continued the process of decarbonization, as the combustion of oil releases only one molecule of carbon for every two of hydrogen. With natural gas (methane), which began its ascent in the final decades of the twentieth century, decarbonization went even further with one unit of carbon being released for every four units of hydrogen. Thus each new major fuel source decreased the carbon/hydrogen ratio. The transition to solar energy will be the final step in this decarbonization process, since renewable energy sources do not release any carbon into the atmosphere.

In previous decades, there was great hope that nuclear power might be the ideal clean fuel to replace coal and oil, but it soon became apparent that nuclear power carries such enormous risks and costs that it is not a viable solution.[78] These risks begin with the contamination of people and the environment with cancer-causing radioactive substances during every stage of the fuel cycle—the mining and enrichment of uranium, the operation and maintenance of the reactor, and the handling and storage or reprocessing of nuclear waste. In addition, there are the unavoidable emissions of radiation in nuclear accidents and even during routine operation of power plants; the unsolved problems of how to safely decommission nuclear reactors and store radioactive waste; the threat of nuclear terrorism and the ensuing loss of basic civic liberties in a totalitarian "plutonium economy"; and the disastrous economic consequences of the use of nuclear power as a capital-intensive, highly centralized source of energy.

All these risks combine with the inherent problems of fuel and construction costs to increase the costs of running nuclear power plants to levels that make them highly uncompetitive. As early as 1977, a prominent utilities investment adviser concluded a thorough investigation of the nuclear industry with the following devastating statement: "The conclusion that must be reached is that, from an economic standpoint alone, to rely upon nuclear fission as the primary source of our stationary energy supplies will constitute economic lunacy on a scale unparalleled in recorded history."[79] Today, nuclear power is the world's slowest growing energy source, dropping to a mere 1 percent growth in 1996, with no prospect of improvement. According to the *Economist*, "Not one [nuclear power plant], anywhere in the world, makes commercial sense."[80]

Solar energy, by contrast, is the energy sector that has seen the fastest growth over the past decade. The use of solar cells (i.e. photovoltaic cells that convert sunlight into electricity) increased by about 17 percent per year in the 1990s, and wind power grew even more spectacularly, by about 24 percent per year.[81] An estimated half a million homes around the world, mostly in remote villages that are not linked to an electric grid, now get their energy from solar cells. The recent invention of solar roofing tiles in Japan promises to lead to a further boost in the use of photovoltaic electricity. As mentioned above, these "solar shingles" are capable of turning rooftops into small power plants, which is likely to revolutionize electricity generation.

These developments show that the transition to solar energy is now well under way. In 1997, a thorough study by five American science laboratories concluded that solar energy could supply 60 percent of the U.S. energy needs at competitive prices today, if there was fair competition and proper accounting of its environmental benefits. One year later, a study by Royal Dutch Shell considered it highly probable that over the next half century, renewable energy sources could become competitive enough to supply at least half of the world's energy needs.[82]

Any long-term solar energy program will have to come up with enough liquid fuel to operate airplanes and at least some of our present ground transportation. Until recently, this has been the Achilles' heel of the solar transition.[83] In the past, the preferred source for a renewable liquid fuel was biomass; in particular, alcohol distilled from fermented grain or fruit. The problem with this solution is that, even though biomass is a renewable resource, the soil in which it grows is not. While we could certainly expect significant alcohol production from special crops, a massive alcohol fuel program would deplete our soil at the same rate as we are now depleting other natural resources.

During the last few years, however, the liquid-fuel problem found a spectacular solution with the development of efficient hydrogen fuel cells that promise to inaugurate a new era in energy production—the "hydrogen economy." Hydrogen, the universe's lightest and most abundant element, is commonly used as rocket fuel. A fuel cell is an electrochemical device that combines hydrogen with oxygen to produce electricity and water—and nothing else! This makes hydrogen the ultimate clean fuel, the definitive last step in the long decarbonization process.

The process in a fuel cell is similar to that in a battery, but uses a continual flow of fuel. Hydrogen molecules are fed into one side of the device, where they are split into protons and electrons by a catalyst. These particles then travel to the other side along different paths. The protons pass through a membrane, while the electrons are forced to travel around it, creating an electric current in the process. After being used, the current reaches the other side of the fuel cell, where the electrons are reunited with the protons and the resulting hydrogen reacts with oxygen from the air to form water. The entire operation is silent, reliable and does not generate any pollution or waste.[84]

Fuel cells were invented in the nineteenth century, but until recently were not produced commercially (except for the U.S. space program), because they were bulky and uneconomical. They required large amounts of platinum as a catalyst, which made them far too expensive for mass production. Besides, fuel cells run on hydrogen, which exists in abundance but must be separated from water (H_2O) or natural gas (CH_4) before it can be used as a fuel. This is not technically difficult, but requires a special infrastructure that nobody in our fossil-fuel economy was interested in developing.

This situation changed radically during the last decade. Technological breakthroughs have drastically reduced the amounts of platinum needed as catalyst, and ingenious "stacking" techniques make it possible to create compact and highly efficient units that will be manufactured within the next few years to supply electricity for our homes, buses, and cars.[85]

While several companies around the world are racing to be the first to produce residential fuel cell systems commercially, a joint venture to create the world's first hydrogen economy was launched by the government of Iceland and several Icelandic corporations.[86] Iceland will use its vast geothermal and hydroelectric resources to produce hydrogen from seawater, to be used in fuel cells first in buses and then in passenger cars and fishing vessels. The goal set by the government is to complete the transition to hydrogen between 2030 and 2040.

At present, natural gas is the most common source of hydrogen, but separation of hydrogen from water with the help of renewable energy sources (especially solar electricity and wind power) will be the most economical—and cleanest—method, in the long run. When that happens, we will have created a truly sustainable system of energy generation. As in nature's ecosystems, all the energy we need will be supplied by the sun, either via small-scale solar devices, or distributed as hydrogen, the ultimate clean fuel, and used in the efficient and reliable operation of fuel cells.

Hypercars

The redesign of automobiles may be the ecodesign branch with the most far-reaching industrial consequences. In typical ecodesign fashion, it began with an analysis of the inefficiency of our present cars, proceeded with a long search for systemic and ecologically oriented solutions, and ended up with design ideas so radical that they will not only change today's automobile industry beyond recognition but may have equally sweeping effects on the associated oil, steel, and electricity industries.

Like many other products of industrial design, the contemporary automobile is stunningly inefficient.[87] Only 20 percent of the energy in the fuel is used to turn the wheels, while 80 percent is lost in the engine's heat and exhaust. Moreover, a full 95 percent of the energy that *is* used moves the car, and only 5 percent moves the driver. The overall efficiency in terms of the proportion of fuel energy used to move the driver is 5 percent of 20 percent—a mere 1 percent!

In the early 1990s, physicist and energy specialist Amory Lovins and his colleagues at the Rocky Mountain Institute took up the challenge of completely redesigning today's vastly inefficient automobile by synthesizing emerging alternative ideas into a conceptual design they called the "hypercar." This design combines three key elements. Hypercars are ultralight, weighing two or three times less than steel cars; they display high aerodynamic efficiency, moving along the road several times more easily than standard cars; and they are propelled by a "hybrid-electric drive," which combines an electric motor with fuel that produces the electricity for the motor on board.

When these three elements are integrated into a single design, they save at least 70 to 80 percent of the fuel used by a standard car, while also making the car safer and more comfortable. In addition, the concept generates numerous surprising effects that promise to revolutionize not only the automobile industry but also industrial design as a whole.[88]

The starting point of the hypercar concept is to reduce the power required to move the vehicle. Since only 20 percent of the fuel energy is used to turn the wheels in a standard car, any saving of power at the wheels will result in a fivefold saving of fuel. In a hypercar, power at the wheels is saved by making the car lighter and more aerodynamic. The standard metal body is replaced by one made of strong carbon fibers embedded in special moldable plastics. Combinations of various fibers offer great design flexibility, and the resulting ultralight body cuts the car's weight in half. In addition, simple streamlining details can reduce air resistance by 40 to 60 percent without restricting stylistic flexibility. Together, these innovations can reduce the power needed to move the car and its passengers by 50 percent or more.

Making the car ultralight generates a cascade of secondary effects, many of which result in further weight reductions. A lighter car can function with a lighter suspension to support the reduced weight, a smaller engine to move it, smaller brakes to decelerate it, and less fuel to run the engine. Moreover, certain components do not merely become smaller but are eliminated altogether. Power steering and power brakes are not needed in ultralight vehicles. The hybrid-electric drive eliminates further components—clutch, transmission, driveshaft etc.—all of which reduce the car's weight.

The new fiber composites are not only ultralight but also extraordinarily strong. They can absorb five times more energy per pound than steel. This is, of course, an important safety element. Hypercars are designed to dissipate crash energy effectively with the help of technologies copied from race cars, which are also ultralight and amazingly safe. In addition to protecting their own occupants, lightweight cars are also less dangerous for the passengers in the vehicles they collide with.

The differences between the physical properties of steel and fiber composites profoundly affect not only the design and operation of hypercars but also their manufacture, distribution, and maintenance. Although carbon fibers are more expensive than steel, the production process of composite car bodies is much more economical. Steel must be pounded, welded, and finished; composites emerge from a mold as a single, finished piece. This cuts tooling costs by up to 90 percent. The car assembly, too, is much simpler, since the lightweight parts are easy to handle and can be lifted without hoists. Painting, which is the most expensive and most polluting step in car manufacture, can be eliminated by integrating color into the molding process.

The multiple advantages of fiber composites combine to favor small design teams, low break-even volumes per model, and local factories, all of which are characteristics of ecodesign as a whole. Maintenance of hypercars is also vastly simpler than that of steel cars, since many of the parts that are frequently responsible for mechanical breakdowns are no longer there. The rust- and fatigue-free composite bodies, which are almost impossible to dent, will last for decades until they are eventually recycled.

Another fundamental innovation is the hybrid-electric drive. Like other electric cars, hypercars have efficient electric motors to turn their wheels, as well as the ability to transform braking energy back into electricity, which offers additional energy savings. Unlike standard electric cars, however, hypercars have no batteries. Instead of using batteries, which continue to be heavy and short-lived, electricity is generated by a small engine, turbine or fuel cell. Such hybrid drive systems are small, and since they are not directly coupled to the wheels, they run near their optimal conditions all the time, which further reduces fuel consumption.

Hybrid cars can use gasoline or a variety of cleaner options, including fuels made from biomass. The cleanest, most efficient and most elegant way to power a hybrid car is to use hydrogen in a fuel cell. Such an automobile not only operates silently and without any pollution, but also becomes, in effect, a small power plant on wheels. This is perhaps the most surprising and far-reaching aspect of the hypercar concept. When the car is parked at the owner's home or place of work—in other words, most of the time—the electricity produced by its fuel cell could be sent into the electric grid and the owner could automatically be credited for it. Amory Lovins estimates that such massive production of electricity would soon put all coal and nuclear power plants out of business, and that a full U.S. fleet of hydrogen-powered hypercars would have five to ten times the generating capacity of the national electric grid, save all the oil OPEC now sells, and reduce America's CO_2 emissions by about two thirds.[89]

When Lovins created the hypercar concept in the early 1990s, he assembled a technical team at his Rocky Mountain Institute to develop the idea. Over the subsequent years the team published numerous professional papers, followed in 1996 by a voluminous report, *Hypercars: Materials, Manufacturing, and Policy Implications.*[90] To maximize competition among car manufacturers, the hypercar team placed all of its ideas in the public domain and shared them conspicuously with some two dozen major car companies.

This unconventional strategy worked as intended, triggering fierce worldwide competition. Toyota and Honda were the first to offer hybrid petrol electric cars—the five-seater Toyota Prius and the two-seater Honda Insight. Similar hybrid cars, achieving fuel efficiencies of 72 to 80 miles per gallon (mpg), were tested by General Motors, Ford and Daimler Chrysler, and are now heading for production. In the meantime, Volkswagen is selling a 78-mpg model in Europe and plans to put a 235-mpg (!) model on the American market in 2003. In addition, fuel-cell cars are slated for production in 2003–05 by eight major automakers.[91]

To increase competitive pressure even further, the Rocky Mountain Institute spun off an independent start-up company, Hypercar Inc., to design the world's first uncompromising, super-efficient and manufacturable hypercar.[92] The design of this concept car was successfully completed in November 2000, and was featured in a frontpage article in *The Wall Street Journal* two months later.[93] It will be a spacious mid-sized sport-utility vehicle (SUV) with fuel efficiency of 99 mpg, which will run silently with zero emissions and a radius of 330 miles, powered by electricity generated in a fuel cell from 7.5 pounds of hydrogen compressed in ultrasafe tanks.[94] The design meets stringent industry standards and is consistent with a 200,000-mile warranty. Lovins and his colleagues hope to produce numerous prototypes by the end of 2002. If they succeed, they will have proven that the hypercar concept can become a commercial reality.

Today, the hypercar revolution is well under way. When the models now in production are in the showrooms of the major car companies, people will buy them not just because they want to save energy and protect the environment, but simply because these new ultralight, safe, pollution-free, silent and super-efficient models will be better cars. People will switch to them just as they switched from mechanical typewriters to computers and from vinyl records to CDs. Eventually, the only steel cars with combustion engines on the road will be a small number of vintage Jaguars, Porsches, Alfa Romeos, and other classic sports cars.

Since the automobile industry is the world's largest, followed by the related oil industry, the hypercar revolution will have a profound impact on industrial production as a whole. Hypercars are an ideal means to introduce examples of the service-and-flow economy advocated by ecodesigners on a large scale. They are likely to be leased rather than sold while the necessary hydrogen infrastructure is developed, and their recyclable materials will flow in closed loops, with toxicities carefully controlled and progressively reduced. The dramatic shifts from steel to carbon fibers and from gasoline to hydrogen will ultimately replace today's steel, petroleum, and related industries with radically different types of environmentally benign and sustainable production processes.

Transition to the Hydrogen Economy

Most of the hybrid cars now in production are not yet powered by fuel cells, since they are still too expensive and hydrogen is not readily available. The production volume needed to bring fuel cell prices down will probably first come from their use in buildings. As mentioned above, there is now fierce worldwide competition for the production of residential fuel cell systems. Until hydrogen fuel can easily be delivered to homes, these systems will include fuel processors that extract hydrogen from natural gas. Thus existing gas lines will be used to provide not only natural gas, but also electricity. Amory Lovins estimates that electricity generated by these fuel cells will easily compete with that from coal and nuclear power plants, because it will not only be produced more cheaply, but will also save the costs of long transmission lines.[95]

Paul Hawken and Amory and Hunter Lovins envisage a scenario for the transition to a hydrogen economy in which the first fuel-cell cars will be leased to people who work in or near buildings with fuel cell systems that extract hydrogen from natural gas.[96] The surplus hydrogen produced by these systems during off-peak hours will be distributed at special filling stations to fuel the hypercars. As the hydrogen market expands with the use of fuel cells in buildings, factories, and vehicles, more centralized production and delivery through new hydrogen pipelines will become attractive.

At first, this hydrogen will also be produced from natural gas, using a special technique that injects the CO_2 resulting from the hydrogen extraction process back into the underground gas fields. In this way, the abundant existing resources of natural gas can be used to produce clean hydrogen fuel without harming the Earth's climate. Eventually, hydrogen will be separated from water with the help of renewable energy from solar cells and wind farms.

As the transition to the hydrogen economy progresses, energy efficiency will outpace oil production so quickly that even cheap oil will become uncompetitive and thus no longer worth extracting. As Amory and Hunter Lovins point out, the Stone Age did not end because people ran out of stones.[97] The Petroleum Age will not end because we will run out of petroleum. It will end because we have developed superior technologies.

Ecodesign Policies

The numerous ecodesign projects reviewed in the preceding pages provide compelling evidence that the transition to a sustainable future is no longer a technical nor a conceptual problem. It is a problem of values and political will. According to the Worldwatch Institute, the policies needed to support ecodesign and the shift to renewable energy include, "a mix of free market competition and regulation, with environmental taxes correcting marketplace distortions; temporary subsidies to support the market entry of renewables; and the removal of hidden subsidies to conventional sources."[98]

The removal of hidden subsidies—or "perverse subsidies," as conservationist Norman Myers calls them[99]—is especially urgent. Today, the governments of the industrial world use vast amounts of their taxpayers' money to subsidize unsustainable and harmful industries and corporate practices. The numerous examples listed by Myers in his eye-opening book, *Perverse Subsidies*, include the billions of dollars paid by Germany to subsidize the extremely harmful coal-burning plants of the Ruhr Valley; the huge subsidies the U.S. government gives to its automobile industry, which was on corporate welfare during most of the twentieth century; the subsidies given to agriculture by the OECD, totalling $300 billion per year, which is paid to farmers to *not* grow food although millions in the world go hungry; as well as the millions of dollars the United States offers to tobacco farmers to grow a crop that causes disease and death.

All of these are perverse subsidies indeed. They are powerful forms of corporate welfare that send distorted signals to the markets. Perverse subsidies are not officially tallied by any government in the world. While they support inequity and environmental degradation, the corresponding life-enhancing and sustainable enterprises are portrayed by the same governments as being uneconomical. It is high time to eliminate these immoral forms of government support.

Another kind of signal the government sends to the marketplace is provided by the taxes it collects. At present, these too are highly distorted. Our existing tax systems place levies on the things we value—jobs, savings, investments—and do not tax the things we recognize as harmful—pollution, environmental degradation, resource depletion, and so on. Like perverse subsidies, this provides investors in the marketplace with inaccurate information about costs. We need to reverse the system: instead of taxing incomes and payrolls, we should tax nonrenewable resources, especially energy, and carbon emissions.[100]

Such a shift in taxation—formerly called "ecological tax reform" and now better known simply as "tax shifting"—would be strictly revenue neutral for the government. This means that taxes would be added to existing products, forms of energy, services and materials, so that their prices would better reflect their true costs, while equal amounts would be subtracted from income and payroll taxes.

To be successful, tax shifting needs to be a slow, long-term process in order to give new technologies and consumption patterns sufficient time to adapt, and it needs to be implemented predictably in order to encourage industrial innovation. Such a long-term, incremental shift of taxation will gradually drive wasteful, harmful technologies and consumption patterns out of the market.

As energy prices go up, with corresponding income tax reductions to offset the increase, people will increasingly switch from conventional to hybrid cars, use bicycles and public transportation, and share carpools when they commute to work. As taxes on petrochemicals and fuel go up, again with offsetting reductions in income taxes, organic farming will become not only the healthiest but also the cheapest means of producing food. Tax shifting will create powerful incentives for business to adopt ecodesign strategies, because all their beneficial effects—increasing resource productivity, reducing pollution, eliminating waste, creating jobs—would also result in tax benefits.

Various forms of tax shifting have recently been initiated in several European nations, including Germany, Italy, the Netherlands, and the Scandinavian countries. Others are likely to follow soon. Indeed, Jacques Delors, former president of the European Commission, is urging governments to adopt the process Europewide. When that happens, the United States will be forced to follow suit so that its businesses remain competitive, because tax shifting will lower their European competitors' labor costs while stimulating innovation.

The taxes people pay in a given society ultimately reflect that society's value system. Hence, a shift to taxation that encourages the creation of jobs, the revitalization of local communities, the conservation of natural resources, and the elimination of pollution reflects the core values of human dignity and ecological sustainability that underlie the principles of ecodesign and the worldwide movement to reshape globalization. As the NGOs in the newly formed global civil society continue to refine their conceptualization of the alternatives to global capitalism and the ecodesign community refines its principles, processes and technologies, tax shifting will be the policy that interlinks and supports both movements, because it reflects the core values they share.

| epilogue |

MAKING SENSE

My objective in this volume has been to develop a conceptual framework that integrates the biological, cognitive, and social dimensions of life; a framework that enables us to adopt a systemic approach to some of the critical issues of our time. The analysis of living systems in terms of four interconnected perspectives—form, matter, process, and meaning—makes it possible to apply a unified understanding of life to phenomena in the realm of matter, as well as to phenomena in the realm of meaning. For example, we saw that metabolic networks in biological systems correspond to networks of communications in social systems; chemical processes producing material structures correspond to thought processes producing semantic structures; and flows of energy and matter correspond to flows of information and ideas.

A central insight of this unified, systemic understanding of life is that its basic pattern of organization is the network. At all levels of life—from the metabolic networks inside cells to the food webs of ecosystems and the networks of communications in human societies—the components of living systems are interlinked in network fashion.

We have seen in particular that in our Information Age, social functions and processes are increasingly organized around networks. Whether we look at corporations, financial markets, the media, or the new global NGOs, we find that networking has become an important social phenomenon and a critical source of power.

As this new century unfolds, there are two developments that will have major impacts on the well-being and ways of life of humanity. Both have to do with networks, and both involve radically new technologies. One is the rise of global capitalism; the other is the creation of sustainable communities based on ecological literacy and the practice of ecodesign. Whereas global capitalism is concerned with electronic networks of financial and informational flows, ecodesign is concerned with ecological networks of energy and material flows. The goal of the global economy is to maximize the wealth and power of its elites; the goal of ecodesign to maximize the sustainability of the web of life.

These two scenarios—each involving complex networks and special advanced technologies—are currently on a collision course. We have seen that the current form of global capitalism is ecologically and socially unsustainable. The so-called "global market" is really a network of machines programmed according to the fundamental principle that money-making should take precedence over human rights, democracy, environmental protection, or any other value.

However, human values can change; they are not natural laws. The same electronic networks of financial and informational flows *could* have other values built into them. The critical issue is not technology, but politics. The great challenge of the twenty-first century will be to change the value system underlying the global economy, so as to make it compatible with the demands of human dignity and ecological sustainability. Indeed, we have seen that this process of reshaping globalization has already begun.

One of the greatest obstacles on the road toward sustainability is the continuing increase in material consumption. In spite of all the emphasis in our new economy on information processing, knowledge generation, and other intangibles, the main goal of these innovations is to increase productivity, which ultimately increases the flow of material goods. Even when Cisco Systems and other Internet companies manage information and expert knowledge without manufacturing any material products, their suppliers and subcontractors do, and many of them, especially in the South, operate with considerable environmental impacts. As Vandana Shiva remarked wryly, "Resources move from the poor to the rich, and pollution moves from the rich to the poor."[1]

Moreover, the software designers, financial analysts, lawyers, investment bankers, and other professionals who have become very wealthy in the "nonmaterial" economy tend to show their wealth by conspicuous consumption. Their large homes, located in sprawling suburbs, are filled with the latest gadgets, their garages stocked with two to three cars per person. Biologist and environmentalist David Suzuki notes that in the last forty years, the size of Canadian families has shrunk by 50 percent, but their living spaces have doubled. "Each person uses four times as much space," Suzuki explains, "because we are all buying so much stuff."[2]

In contemporary capitalist society, the central value of money-making goes hand in hand with the glorification of material consumption. A never-ending stream of advertising messages reinforces people's delusion that the accumulation of material goods is the royal road to happiness, the very purpose of our lives.[3] The United States projects its tremendous power around the world to maintain optimal conditions for the perpetuation and expansion of production. The central goal of its vast empire—its overwhelming military might, impressive range of intelligence agencies, and dominant positions in science, technology, media, and entertainment—is not to expand its territory, nor to promote freedom and democracy, but to make sure that it has global access to natural resources and that markets around the world remain open to its products.[4] Accordingly, political rhetoric in America moves swiftly from "freedom" to "free trade" and "free markets." The free flow of capital and goods is equated with the lofty ideal of human freedom, and material acquisition is portrayed as a basic human right, increasingly even as an obligation.

This glorification of material consumption has deep ideological roots that go far beyond economics and politics. Its origins seem to lie in the universal association of manhood with material possessions in patriarchal cultures. Anthropologist David Gilmore studied images of manhood around the world—"male ideologies," as he puts it—and found striking cross-cultural similarities.[5] There is a recurring notion that "real manhood" is different from simple biological maleness, that it is something that has to be won. In most cultures, Gilmore shows, boys "must earn the right" to be called men. Although women, too, are judged by sexual standards that are often stringent, Gilmore notes that their very status as women is rarely questioned.[6]

In addition to well-known images of manliness like physical strength, toughness, and aggression, Gilmore found that in culture after culture, "real" men have traditionally been those who produce more than they consume. The author emphasizes that the ancient association of manhood with material production meant production on behalf of the community: "Again and again we find that 'real' men are those who give more than they take; they serve others. Real men are generous, even to a fault."[7]

Over time, there was a shift in this image, from production for the sake of others to material possession for the sake of one's self. Manhood was now measured in terms of ownership of valuable goods—land, cattle, or cash—and in terms of power over others, especially women and children. This image was reinforced by the universal association of virility with "bigness"—as measured in muscle strength, accomplishments, or number of possessions. In modern society, Gilmore points out, male "bigness" is measured increasingly by material wealth: "The Big Man in any industrial society is also the richest guy on the block, the most successful, the most competent . . . He has the most of what society needs or wants."[8]

The association of manhood with the accumulation of possessions fits well with other values that are favored and rewarded in patriarchal culture—expansion, competition, and an "object-centered" consciousness. In traditional Chinese culture, these were called *yang* values and were associated with the masculine side of human nature.[9] They were not seen as being intrinsically good or bad. However, according to Chinese wisdom, the *yang* values need to be balanced by their *yin*, or feminine, counterparts—expansion by conservation, competition by cooperation, and the focus on objects by a focus on relationships. I have long argued that the movement toward such a balance is very consistent with the shift from mechanistic to systemic and ecological thinking that is characteristic of our time.[10]

Among the many grassroots movements working for social change today, the feminist movement and the ecology movement advocate the most profound value shifts, the former through a redefinition of gender relationships, the latter through a redefinition of the relationship between humans and nature. Both can contribute significantly to overcoming our obsession with material consumption.

By challenging the patriarchal order and value system, the women's movement has introduced a new understanding of masculinity and personhood that does not need to associate manhood with material possessions. At the deepest level, feminist awareness is based on women's experiential knowledge that all life is connected, that our existence is always embedded in the cyclical processes of nature.[11] Feminist consciousness, accordingly, focuses on finding fulfillment in nurturing relationships rather than in the accumulation of material goods.

The ecology movement arrives at the same position from a different approach. Ecological literacy requires systemic thinking—thinking in terms of relationships, context, patterns, and processes—and ecodesigners advocate the transition from an economy of goods to an economy of service and flow. In such an economy, matter cycles continually, so that the net consumption of raw materials is drastically reduced. As we have seen, a service-and-flow or zero-emissions economy is also excellent for business. As wastes turn into resources, new revenue streams are generated, new products are created and productivity increases. Whereas the extraction of resources and the accumulation of waste are bound to reach their ecological limits, the evolution of life has demonstrated for more than three billion years that in a sustainable

Earth household, there are no limits to development, diversification, innovation, and creativity.

In addition to increasing resource productivity and reducing pollution, the zero-emissions economy also increases employment opportunities and revitalizes local communities. Thus the rise of feminist awareness and the movement toward ecological sustainability will combine to bring about a profound change of thinking and values—from linear systems of resource extraction and accumulation of products and waste to cyclical flows of matter and energy; from the focus on objects and natural resources to a focus on services and human resources; from seeking happiness in material possessions to finding it in nurturing relationships. In the eloquent words of David Suzuki:

> Family, friends, community—these are the sources of the greatest love and joy we experience as humans. We visit family members, keep in touch with favourite teachers, share and exchange pleasantries with friends. We undertake difficult projects to help others, save frogs or protect a wilderness, and in the process discover extreme satisfaction. We find spiritual fulfillment in nature or by helping others. None of these pleasures requires us to consume things from the Earth, yet each is deeply fulfilling. These are complex pleasures, and they bring us much closer to real happiness than the simple ones, like a bottle of Coke or a new minivan.[12]

The question naturally arises: Will there be enough time for this profound change of values to halt and reverse the present depletion of natural resources, extinction of species, pollution, and global climate change? The developments mentioned in the preceding pages do not point to a clear answer. If we extrapolate current environmental trends into the future, the outlook is alarming. On the other hand, there are many signs that a significant, and perhaps decisive, number of people and institutions around the world have begun the transition to ecological sustainability. Many of my colleagues in the ecology movement share this view, as the following three voices, representative of many others, make clear.[13]

> I believe that there are now some clear signs that the world does seem to be approaching a kind of paradigm shift in environmental consciousness. Across a spectrum of activities, places, and institutions, the atmosphere has changed markedly in just the last few years.
>
> *Lester Brown*

> I am more hopeful now than I was a few years ago. I think the speed and importance of things getting better outweighs the speed and importance of things getting worse. One of the most hopeful developments is the co-operation between the North and the South in the global civil society. We have much richer expertise available now than we had before.
>
> *Amory Lovins*

> I am optimistic, because life has its own ways of not becoming extinct; and people, too, have their own ways. They will continue life's tradition.
>
> *Vandana Shiva*

To be sure, the transition to a sustainable world will not be easy. Gradual changes will not be enough to turn the tide; we also need some major breakthroughs. The task seems overwhelming, but is not impossible. From our new understanding of complex biological and social systems we have learned that meaningful disturbances can trigger multiple feedback processes that may rapidly lead to the emergence of new order. Recent history has shown us some powerful examples of these dramatic transformations—from the fall of the Berlin Wall and the Velvet Revolution in Europe to the end of Apartheid in South Africa.

On the other hand, complexity theory also tells us that these points of instability may lead to breakdowns rather than breakthroughs. So what can we hope for the future of humanity? In my opinion, the most inspiring answer to this existential question comes from one of the key figures in the recent dramatic social transformations, the great Czech playwright and statesman Václav Havel, who turns the question into a meditation on hope itself:

The kind of hope that I often think about . . . I understand above all as a state of mind, not a state of the world. Either we have hope within us or we don't; it is a dimension of the soul, and it's not essentially dependent on some particular observation of the world or estimate of the situation . . . [Hope] is not the conviction that something will turn out well, but the certainty that something makes sense, regardless of how it turns out.[14]

| notes |

Preface

1. The statement that serves as the motto for this book was made by Czech President Václav Havel in his opening address to the Forum 2000 Conference in Prague on 15 October 2000.

One: The Nature of Life

1. The following account has been inspired by Luisi (1993) and by stimulating correspondence and discussions with the author.

2. See Capra (1996), pp. 257ff.; see also pp. 58ff.

3. See pp. 16–17.

4. Some parts of cells, such as mitochondria and chloroplasts, were once independent bacteria that invaded larger cells and then coevolved with them to form new composite organisms; see Capra (1996), p. 231. These organelles still reproduce at different times from the rest of the cell, but they cannot do so without the functioning of the integrated cell and can therefore no longer be considered autonomous living systems; see Morowitz (1992), p. 231.

5. See Morowitz (1992), pp. 59ff.

6. Ibid., pp. 66ff.

7. Ibid., p. 54.

8. See Lovelock (1991); Capra (1996), pp. 100ff.

9. Morowitz (1992), p. 6.

10. See *New York Times*, 11 July 1997.

11. Luisi (1993).

12. See pp. 21–22.

13. Margulis, personal communication, 1998.

14. See, for example, Capra (1996), p. 165.

15. Margulis, personal communication, 1998.

16. See Capra (1996), p. 280.

17. Margulis (1998a), p. 63.

18. Excluded from this production are the primary components such as oxygen, water, and CO_2, as well as the food molecules entering the cell.

19. See Capra (1996), pp. 97ff.

20. See Luisi (1993).

21. Ibid.

22. Ibid.

23. See Morowitz (1992), p. 99.

24. See Capra (1996), p. 165.

25. See Capra (1996), p. 132.

26. Goodwin (1994), Stewart (1998).

27. Stewart (1998), p. xii.

28. See p. 171 for a more extensive discussion of genetic determinism.

29. Margulis, personal communication, 1998.

30. See Capra (1996), pp. 86ff.

31. It is interesting to note that "complexity" is derived etymologically from the Latin verb *complecti* ("to twine together") and the noun *complexus* ("network"). Thus, the idea of nonlinearity—a network of intertwined strands—lies at the very root of the meaning of "complexity."

32. Brian Goodwin, personal communication, 1998.

33. See Capra (1996), p. 86.

34. See Margulis and Sagan (1995), p. 57.

35. Luisi (1993).

36. See Capra (1996), pp. 92–94.

37. See Gesteland, Cech, and Atkins (1999).

38. See Gilbert (1986).

39. Szostak, Bartel, and Luisi (2001).

40. Luisi (1998).

41. Morowitz (1992).

42. Ibid., p. 154.

43. Ibid., p. 44.

44. See ibid., pp. 107–08.

45. Ibid., pp. 174–75.

46. Ibid., pp. 92–93.

47. See p. 29.

48. See Morowitz (1992), p. 154.

49. Ibid., p. 9.

50. Ibid., p. 96.

51. Luisi (1993 and 1996).

52. See Fischer, Oberholzer, and Luisi (2000).

53. See Morowitz (1992), pp. 176–77.

54. Pier Luigi Luisi, personal communication, January 2000.

55. See Capra (1996), pp. 88–89, 92ff.

56. Morowitz (1992), p. 171.
57. See ibid., pp. 119ff.
58. Ibid., pp. 137, 171.
59. Ibid., p. 88.
60. See Capra (1996), pp. 228ff.
61. However, very recent research in genetics seems to indicate that the rate of mutation is not a matter of pure chance, but is regulated by the cell's epigenetic network; see pp. 166–67.
62. Margulis (1998b).
63. Margulis, personal communication, 1998.
64. See Sonea and Panisset (1993).
65. See Capra (1996), pp. 230ff.
66. See Margulis (1998a), pp. 45ff.
67. Margulis and Sagan (1997).
68. See Gould (1994).
69. Margulis (1998a), p. 8.

Two: Mind and Consciousness

1. Revonsuo and Kamppinen (1994), p. 5.
2. See Capra (1996), pp. 96–97 and 173–74.
3. See ibid., pp. 266ff.
4. See Capra (1982), pp. 169–70.
5. See Varela (1996a), Tononi and Edelman (1998).
6. See, e.g., Crick (1994), Dennett (1991), Edelman (1989), Penrose (1994); *Journal of Consciousness Studies*, vols. 1–6, 1994–99; Tucson II Conference, "Toward a Science of Consciousness," Tucson, Arizona, 13–17 April 1996.
7. See Edelman (1992), pp. 122–23.
8. See ibid., p. 112.
9. See Searle (1995).
10. See Chalmers (1995).
11. See Capra (1996), pp. 24ff.
12. Varela (1999).
13. See Varela and Shear (1999).
14. See ibid.
15. See Varela (1996a).
16. See Churchland and Sejnowski (1992), Crick (1994).
17. Crick (1994), p. 3.
18. Searle (1995).
19. See ibid., Varela (1996a).
20. Dennett (1991).
21. See Edelman (1992), pp. 220ff.
22. See McGinn (1999).

23. Varela (1996a).
24. Capra (1988), p. 138.
25. *Journal of Consciousness Studies*, vol. 6, no. 2–3, 1999.
26. See Vermersch (1999).
27. See ibid.
28. See Varela (1996a), Depraz (1999).
29. See Shear and Jevning (1999).
30. See Wallace (1999).
31. See Varela et al. (1991), Shear and Jevning (1999).
32. Penrose (1999); see also Penrose (1994).
33. Edelman (1992), p. 211.
34. See, e.g., Searle (1984), Edelman (1992), Searle (1995), Varela (1996a).
35. Varela (1995), Tononi and Edelman (1998).
36. Tononi and Edelman (1998).
37. See Varela (1995); see also Capra (1996), pp. 292–93.
38. See Varela (1996b).
39. See Varela (1996a), Varela (1999).
40. See Tononi and Edelman (1998).
41. See Edelman (1989), Edelman (1992).
42. See p. 38; see also Capra (1996), pp. 257ff.
43. Núñez (1997).
44. Maturana (1970), Maturana and Varela (1987), pp. 205ff.; see also Capra (1996), pp. 287ff.
45. See p. 35.
46. See Maturana (1995).
47. Maturana (1998).
48. Maturana and Varela (1987), p. 245.
49. Fouts (1997).
50. Ibid., p. 57.
51. See Wilson and Reeder (1993).
52. See Fouts (1997), p. 365.
53. Ibid., p. 85.
54. See ibid., pp. 74ff.
55. Ibid., pp. 72, 88.
56. Ibid., pp. 302–03.
57. See ibid., p. 191.
58. Kimura (1976); see also Iverson and Thelen (1999).
59. Fouts (1997), pp. 190–91.
60. See ibid., pp. 193–95.
61. See ibid., pp. 184ff.
62. Ibid., p. 192.
63. Ibid., p. 197.
64. See Johnson (1987), Lakoff (1987), Varela et al. (1991), Lakoff and Johnson (1999).

65. Lakoff and Johnson (1999).
66. Ibid., p. 4.
67. See Lakoff (1987).
68. See ibid., pp. 24ff.
69. Lakoff and Johnson (1999), pp. 34–35.
70. See ibid., pp. 380–81.
71. See ibid., pp. 45ff.
72. See ibid., p. 46.
73. See ibid., pp. 60ff.
74. Ibid., p. 3.
75. Ibid., p. 551.
76. Searle (1995).
77. Lakoff and Johnson (1999), p. 4.
78. See pp. 9–10.
79. See pp. 36–37.
80. Steindl-Rast (1990).
81. See Capra and Steindl-Rast (1991), pp. 14–15.

Three: Social Reality

1. See Capra (1996), pp. 157ff.

2. The emergence and refinement of the concept of *pattern of organization* has been a crucial element in the development of systems thinking. Maturana and Varela, in their theory of autopoiesis, distinguish clearly between the *organization* and the *structure* of a living system; and Prigogine coined the term "dissipative *structure*" to characterize the physics and chemistry of open systems that operate far from equilibrium. See Capra (1996), pp. 17ff., 98, 88–89.

3. See pp. 9–10.

4. See Searle (1984), p. 79.

5. I am grateful to Otto Scharmer for pointing this out to me.

6. See, for example, Windelband (1901), pp. 139ff.

7. For a concise review of twentieth-century social theory, see Baert (1998), on which the following pages are largely based.

8. See p. 82.

9. See Baert (1998), pp. 92ff.

10. See ibid., 103–04.

11. Ibid., pp. 134ff.

12. See, e.g., Held (1990).

13. See Capra (1996), pp. 211–12.

14. See Luhmann (1990); see also Medd (2000) for an extensive review of Luhmann's theory.

15. See p. 108.

16. Luhmann (1990).

17. See Searle (1984), pp. 95ff.

18. See p. 35.

19. See Williams (1981).

20. Galbraith (1984); portions reprinted as "Power and Organization" in Lukes (1986).

21. See ref. 20. Instead of "coercive" Galbraith actually uses the arcane "condign," which means "appropriate" and is used mostly in reference to punishment.

22. See David Steindl-Rast in Capra and Steindl-Rast (1991), p. 190.

23. Galbraith, ref. 20.

24. Quoted in Lukes (1986), p. 28.

25. Ibid., p. 62.

26. The complex interactions between formal organizational structures and informal networks of communications, which exist within all organizations, are discussed in some detail below; see pp. 110–11.

27. Castells, personal communication, 1999.

28. See p. 59.

29. See p. 35.

30. See, for example, Fischer (1985).

31. Castells (2000b); for references to similar definitions by Harvey Brooks and Daniel Bell, see Castells (1996), p. 30.

32. See p. 58.

33. See Capra (1996), p. 29.

34. See Kranzberg and Pursell (1967).

35. See Morgan (1998), pp. 270ff.

36. See Ellul (1964), Winner (1977), Mander (1991), Postman (1992).

37. Kranzberg and Pursell (1967), p. 11.

Four: Life and Leadership in Organizations

1. See p. 242.

2. See Wheatley and Kellner-Rogers (1998).

3. My understanding of the nature of human organizations and the relevance of the systems view of life to organizational change has been shaped decisively by extensive collaboration with Margaret Wheatley and Myron Kellner-Rogers, with whom I conducted a series of seminars on self-organizing systems at Sundance, Utah, during 1996–97.

4. See p. 10.

5. Wheatley and Kellner-Rogers (1998).

6. See Castells (1996), p. 17; see also p. 114.

7. See Chawla and Renesch (1995), Nonaka and Takeuchi (1995), Davenport and Prusak (2000).

8. See pp. 12 and 35.

9. See p. 88.

10. See de Geus (1997a), p. 154.

11. Block (1993), p. 5.

12. Morgan (1998), p. xi.

13. See Capra (1982); Capra (1996), pp. 19ff.

14. See Morgan (1998), pp. 21ff.

15. Morgan (1998), pp. 27–28.

16. Senge (1996); see also Senge (1990).

17. Senge (1996).

18. Ibid.

19. De Geus (1997a).

20. See ibid., p. 9.

21. Ibid., p. 21.

22. Ibid., p. 18. It is a great pity that Shell, apparently, paid very little attention to this exhortation from one of its top executives. After its environmentally disastrous oil extraction in Nigeria during the early 1990s and the subsequent tragic execution of Ken Saro-Wiwa and eight other Ogoni freedom fighters, an independent investigation took place, headed by Professor Claude Aké, director of Nigeria's Center for Advanced Social Studies. According to Aké, Shell continued to display the insensitive and arrogant attitude that is typical of multinational oil companies. He was baffled, Aké said, by the corporate culture of the oil companies. "Frankly," he mused, "I would have expected a much more sophisticated corporate strategy from Shell" (*Manchester Guardian Weekly*, 17 December 1995).

23. See p. 82.

24. See *Business Week*, 13 September 1999.

25. See Cohen and Rai (2000).

26. See p. 216.

27. See Wellman (1999).

28. Castells (1996); see also p. 131.

29. Wenger (1996).

30. Wenger (1998), pp. 72ff.

31. See p. 85.

32. De Geus (1997b).

33. Wenger (1998), p. 6.

34. I am grateful to Angelika Siegmund for extended discussions of this topic.

35. It should be noted, however, that not all informal networks are fluid and self-generating. For example, the well-known "old boys" networks are informal patriarchal structures that can be very rigid and may exert considerable power. When I speak of "informal structures" in the following paragraphs, I refer to continually self-generating networks of communications, or communities of practice.

36. See Wheatley and Kellner-Rogers (1998).

37. See p. 36.

38. Wheatley and Kellner-Rogers (1998).

39. See Capra (1996), pp. 34–35.

40. See p. 88.

41. Tuomi (1999).

42. See Nonaka and Takeuchi (1995).
43. Nonaka and Takeuchi (1995), p. 59.
44. See Tuomi (1999), pp. 323ff.
45. See Winograd and Flores (1991), pp. 107ff.
46. See p. 51.
47. Wheatley (2001).
48. Wheatley (1997).
49. See p. 13.
50. Quoted in Capra (1988), p. 20.
51. See Capra (1975).
52. Proust (1921).
53. See p. 91.
54. See Capra (2000).
55. See p. 65.
56. See pp. 72–73.
57. I am grateful to Morten Flatau for extensive discussions of this point.
58. Wheatley (1997).
59. See p. 64.
60. Wheatley and Kellner-Rogers (1998).
61. De Geus (1997b).
62. Siegmund, personal communication, July 2000.
63. De Geus (1997a), p. 57.
64. See *The Economist*, 22 July 2000.
65. See, for example, Petzinger (1999).
66. See Castells (1996); see also p. 136.

Five: The Networks of Global Capitalism

1. Mander and Goldsmith (1996).
2. Castells (1996).
3. Ibid., p. 4.
4. Castells (1996–98).
5. Giddens (1996).
6. See Castells (1998), pp. 4ff.
7. Ibid., p. 338.
8. Hutton and Giddens (2000).
9. Václav Havel, remarks during discussions at Forum 2000, 10–13 October 1999.
10. See p. 119.
11. See Castells (1996), pp. 40ff.
12. See Capra (1996), pp. 51ff.
13. See Abbate (1999).
14. See Himanen (2001).
15. See Capra (1982), pp. 211ff.

16. See Castells (1996), pp. 18–22; Castells (2000a).
17. Castells (1996), pp. 434–35.
18. Castells (1998), p. 341.
19. Giddens in Hutton and Giddens (2000), p. 10.
20. See Castells (2000a).
21. Ibid.
22. See Volcker (2000).
23. See Faux and Mishel (2000).
24. Volcker (2000).
25. Castells, personal communication, 2000.
26. Kuttner (2000).
27. Castells (2000a).
28. See p. 214.
29. See p. 126.
30. See Castells (1996), pp. 474–75.
31. Castells (1996), p. 476.
32. See Castells (1998), pp. 70ff.
33. UNDP (1996).
34. See UNDP (1999).
35. See Castells (1998), pp. 130–31.
36. See Castells (2000a).
37. Castells (1998), p. 74.
38. See ibid., pp. 164–65.
39. See Capra (1982), p. 225.
40. See Brown et al. (2001) and preceding annual reports; see also Gore (1992), Hawken (1993).
41. Gore (1992).
42. Goldsmith (1996).
43. See ibid.
44. See Shiva (2000).
45. Ibid.
46. Goldsmith (1996).
47. Ibid.
48. See Castells (1996), pp. 469ff.
49. See Castells (1998), pp. 346–47.
50. The same can be said about the new phenomenon of international terrorism, as the attacks against the United States on 11 September 2001 have dramatically shown; see Zunes (2001).
51. Ibid., pp. 166ff.
52. Ibid., p. 174.
53. Ibid., pp. 179–80.
54. Ibid., pp. 330ff.
55. Ibid., p. 330.
56. See Korten (1995) and Korten (1999).

57. Manuel Castells, personal communication, 1999.
58. See Capra (1982), pp. 279–80.
59. See Capra (1996), p. 35.
60. See Castells (1996), pp. 327ff.
61. See p. 85.
62. Castells (1996), p. 329.
63. McLuhan (1964).
64. See Danner (2000).
65. See Castells (1996), p. 334.
66. See p. 111.
67. See Castells (1996), pp. 339–40.
68. Castells, personal communication, 1999.
69. See Schiller (2000).
70. See p. 53.
71. Castells (1996), p. 371.
72. See ibid., p. 476.
73. Castells (1998), p. 348.
74. George Soros, remarks at Forum 2000, Prague, October 1999; see also Soros (1998).
75. Castells (2000a).
76. See p. 226.

Six: Biotechnology at a Turning Point

1. See p. 12.
2. Keller (2000).
3. Ho (1998a), p. 19; see also Holdrege (1996) for an eminently readable introduction to genetics and genetic engineering.
4. See Capra (1982), pp. 116ff.
5. See Ho (1998a), pp. 42ff.
6. See Margulis and Sagan (1986), pp. 89–90.
7. Ho (1998a), pp. 146ff.
8. See *Science*, 6 June 1975; pp. 991ff.
9. Although these animals were created through genetic manipulation rather than sexual reproduction, they are not clones in the strict sense of the term; see p. 184.
10. See Altieri (2000b).
11. See p. 197.
12. Ho (1998a), pp. 14ff.
13. See *New York Times*, 13 February 2001.
14. See ibid.
15. *Nature*, 15 February 2001; *Science*, 16 February 2001.
16. Keller (2000), p. 138.
17. Bailey, quoted in Keller (2000), pp. 129–30.

18. A gene consists of a sequence of elements, called "nucleotides," along a strand of the DNA double helix; see, for example, Holdrege (1996), p. 74.
19. Keller (2000), p. 14.
20. See ibid., pp. 26ff.
21. See ibid., p. 27.
22. Ibid., p. 31.
23. See ibid., pp. 32ff.
24. Ibid., p. 34.
25. See Capra (1996), pp. 224–25.
26. Shapiro (1999).
27. See p. 30.
28. See p. 34.
29. McClintock (1983).
30. See Watson (1968).
31. Quoted in Keller (2000), p. 54.
32. Ho (1998a), p. 99.
33. Strohman (1997).
34. See Keller (2000), pp. 59ff.
35. See Baltimore (2001).
36. See Keller (2000), p. 61.
37. Ibid., p. 63.
38. See ibid., pp. 64ff.
39. Ibid., p. 57.
40. See ibid., p. 100.
41. See ibid., pp. 55ff.
42. See ibid., pp. 90ff.
43. See Strohman (1997).
44. See, for example, Kauffman (1995), Stewart (1998), Solé and Goodwin (2000).
45. See Capra (1996), p. 26.
46. See Keller (2000), pp. 112–13.
47. Ibid., pp. 103ff.
48. See ibid., pp. 111ff.
49. Dawkins (1976).
50. Keller (2000), p. 115; see also Goodwin (1994), pp. 29ff., for a critical discussion of the "selfish gene" metaphor.
51. I am grateful to Brian Goodwin for clarifying discussions on this subject.
52. See Capra (1996), pp. 128ff., for a brief introduction to the mathematical language of complexity theory.
53. Gelbart (1998).
54. Keller (2000), p. 9.
55. Holdrege (1996), pp. 116–17.
56. See ibid., pp. 109ff.
57. Ehrenfeld (1997).
58. Strohman (1997).

59. Weatherall (1998).
60. See Lander and Schork (1994).
61. See Ho (1998a), p. 190.
62. Keller (2000), p. 68.
63. Strohman (1997).
64. Ho (1998a), p. 35.
65. In its strict sense, the term "clone" refers to one or several organisms derived from a single parent by asexual reproduction, as in a pure culture of bacteria. Except for differences due to mutations, all members of a clone are genetically identical to the parent.
66. Lewontin (1997).
67. Ibid.
68. See Ho (1998a), pp. 174–75.
69. For example, the cellular structures known as mitochondria (the cell's "powerhouses") contain their own genetic material and reproduce independently from the rest of the cell; see Capra (1996), p. 231. Their genes are involved in the production of some essential enzymes.
70. See Lewontin (1997).
71. See Ho (1998a), p. 179.
72. See ibid., pp. 180–81.
73. See Capra (1982), pp. 253ff.
74. Ehrenfeld (1997).
75. See Altieri and Rosset (1999).
76. See Simms (1999).
77. See *Guardian Weekly*, p. 13 June 1999.
78. See ibid.
79. Altieri and Rosset (1999).
80. Lappé, Collins, and Rosset (1998).
81. See Simms (1999).
82. Altieri (2000a).
83. See Altieri and Rosset (1999).
84. Simms (1999).
85. See Jackson (1985), Altieri (1995); see also Mollison (1991).
86. See Capra (1996), pp. 298ff.
87. See Hawken, Lovins, and Lovins (1999), p. 205.
88. See Norberg-Hodge, Merrifield, and Gorelick (2000).
89. See Halweil (2000).
90. See Altieri and Uphoff (1999); see also Pretty and Hine (2000).
91. Quoted in Altieri and Uphoff (1999).
92. Ibid.
93. Altieri (2000a).
94. See Altieri (2000b).
95. See pp. 159–60.
96. Bardocz (2001).

97. Meadows (1999).
98. See Altieri (2000b).
99. See Shiva (2000).
100. See Shiva (2001).
101. See Steinbrecher (1998).
102. See Altieri (2000b).
103. Losey et al. (1999).
104. See Altieri (2000b).
105. See Ho (1998b), Altieri (2000b).
106. Stanley et al. (1999).
107. Ehrenfeld (1997).
108. See Altieri and Rosset (1999).
109. Shiva (2000).
110. See ibid.
111. See p. 187.
112. See Mooney (1988).
113. See Ho (1998a), p. 26.
114. See Shiva (1997).
115. Shiva (2000).
116. See p. 226.
117. See Ho (1998a), pp. 246ff.; Simms (1999).
118. See p. 232.
119. Benyus (1997).
120. Strohman (1997).
121. See p. 176.

Seven: Changing the Game

1. See Brown et al. (2001).
2. See Hawken, Lovins and Lovins (1999), p. 3.
3. Quoted in Brown et al. (2001), p. 10; see also McKibben (2001).
4. See ibid., pp. xvii–xviii and pp. 10ff.
5. See *New York Times*, 19 August 2000.
6. See Brown et al. (2001), p. 10.
7. See Capra (1982), p. 277.
8. See Brown et al. (2001), p. xviii and pp. 10–11.
9. See ibid., pp. 123ff.
10. See ibid., p. 137.
11. Janet Abramovitz in Brown et al. (2001), pp. 123–24.
12. See Brown et al. (2001), pp. 4–5.
13. See p. 156.
14. See p. 135.
15. See Castells (2000a).

16. See Barker and Mander (1999), Wallach and Sforza (2001).
17. See p. 147.
18. See Henderson (1999), pp. 35ff.
19. See *Guardian Weekly*, 1–7 February 2001.
20. See p. 129.
21. See Capra and Steindl-Rast (1991), pp. 16–17.
22. See Union of International Associations, www.uia.org; see also Union of International Associations (2000/2001).
23. See, e.g., Barker and Mander (1999).
24. See Hawken (2000).
25. Hawken (2000).
26. Quoted in ibid.
27. See Khor (1999/2000).
28. See Global Trade Watch, www.tradewatch.org.
29. *Guardian Weekly*, 8–14 February 2001.
30. See p. 148.
31. Castells (1997), pp. 354ff.
32. See p. 132.
33. Warkentin and Mingst (2000).
34. Quoted in Warkentin and Mingst (2000).
35. It is interesting to note that this new form of political discourse was introduced by the German Greens in the early 1980s when they first came to power; see Capra and Spretnak (1984), p. xiv.
36. See p. 156.
37. Warkentin and Mingst (2000).
38. Castells (1998), pp. 352–53.
39. Debi Barker, IFG, personal communication, October 2001.
40. See p. 107 and p. 151.
41. Robbins (2001), p. 380.
42. See, e.g., "The Monsanto Files," special issue of *The Ecologist*, September/October 1998.
43. Robbins (2001), pp. 372ff.; see also Tokar (2001).
44. See Robbins (2001), p. 374.
45. *Wall Street Journal* 7 January 2000.
46. Brown (1981).
47. World Commission on Environment and Development (1987).
48. See p. 214.
49. See Orr (1992); Capra (1996), pp. 297ff.; Callenbach (1998).
50. See Barlow and Crabtree (2000).
51. Benyus (1997), p. 2.
52. See p. 120.
53. See Hawken (1993), McDonough and Braungart (1998).
54. See Pauli (1996).
55. See Pauli (2000); see also ZERI website, www.zeri.org.

56. See p. 141.

57. See ZERI website, www.zeri.org.

58. McDonough and Braungart (1998).

59. Ibid.

60. See Brown (1999).

61. See Hawken, Lovins and Lovins (1999), pp. 185–86.

62. Hawken (1993), p. 68.

63. See McDonough and Braungart (1998); see also Hawken, Lovins and Lovins (1999), pp. 16ff.

64. See Anderson (1998); see also Hawken, Lovins and Lovins (1999), 139–41.

65. See Canon website, www.canon.com.

66. See website of the Fiat Group, www.fiatgroup.com.

67. See Hawken, Lovins and Lovins (1999), pp. 11–12.

68. See Gardner and Sampat (1998).

69. Hawken, Lovins and Lovins (1999), pp. 10–12.

70. See ibid., pp. 94ff.

71. McDonough and Braungart (1998).

72. See Hawken, Lovins and Lovins (1999), pp. 94, 102–103; see also Orr (2001).

73. See p. 147 above.

74. See Register and Peeks (1997), Register (2001).

75. Newman and Kenworthy (1998); see also Jeff Kenworthy, "City Building and Transportation Around the World," in Register and Peeks (1997).

76. See p. 254.

77. Dunn (2001).

78. See Capra (1982), pp. 242ff.

79. Quoted ibid., p. 400.

80. Quoted in Hawken, Lovins and Lovins (1999), p. 249.

81. See Dunn (2001).

82. See Hawken, Lovins and Lovins (1999), pp. 247–48.

83. See Capra (1982), pp. 403ff.

84. See "The Future of Fuel Cells," Special Report, *Scientific American*, July 1999.

85. See Lamb (1999), Dunn (2001).

86. See Dunn (2001).

87. See Hawken, Lovins and Lovins (1999), p. 24.

88. See ibid., pp. 22ff.

89. Ibid., pp. 35–37. Independence from OPEC oil would enable the United States to change radically its foreign policy in the Middle East, which is presently driven by the perceived need for oil as a "strategic resource." A shift from such a resource-oriented policy would significantly change the conditions underlying the recent wave of international terrorism. Hence, an energy policy based on renewable energy sources and conservation is not only imperative for moving toward ecological sustainability, but must also be seen as vital to America's national security; see Capra (2001).

90. Lovins et al. (1996).

91. See Lovins and Lovins (2001).

92. See www.hypercar.com.

93. *The Wall Street Journal*, 9 January 2001.

94. See Denner and Evans (2001).

95. See Hawken, Lovins and Lovins (1999), p. 34.

96. Ibid., pp. 36–37.

97. Lovins and Lovins (2001).

98. Dunn (2001).

99. Myers (1998).

100. See Hawken (1993), pp. 169ff.; Daly (1995).

Epilogue: Making Sense

1. Vandana Shiva, quoted on p. 147.

2. Suzuki (2001).

3. See Dominguez and Robin (1999).

4. See Ramonet (2000).

5. Gilmore (1990).

6. Curiously, Gilmore does not mention the fact, widely discussed in feminist literature, that there is no need for women to prove their womanhood, because of their ability to give birth, which was perceived as an awesome, transformative power in prepatriarchal cultures; see, for example, Rich (1977).

7. Gilmore (1990), p. 229. However, psychologist Vera van Aaken points out that in patriarchal cultures the definition of manhood in terms of warrior qualities takes priority over that in terms of generous material production, and that Gilmore tends to downplay the suffering inflicted on the community by the warrior ideal; see van Aaken (2000), p. 149.

8. Gilmore (1990), p. 110.

9. See Capra (1982), pp. 36ff.

10. See Capra (1996), pp. 3ff.

11. See Spretnak (1981).

12. Suzuki and Dressel (1999), pp. 263–64.

13. Brown (1999); Lovins, personal communication, May 2001; Shiva, personal communication, February 2001.

14. Havel (1990), p. 181.

| bibliography |

Aaken, Vera Van. *Männliche Gewalt [Male Violence]*. Patmos, Düsseldorf, Germany, 2000.

Abbate, Janet. *Inventing the Internet*. MIT Press, 1999.

Altieri, Miguel. *Agroecology*. Westview Press, Boulder, Colo., 1995.

———. "Biotech Will Not Feed the World." *San Francisco Chronicle*, 30 March 2000a.

———. "The Ecological Impacts of Transgenic Crops on Agroecosystem Health." *Ecosystem Health*, vol. 6, no. 1, March 2000b.

———, and Peter Rosset. "Ten Reasons Why Biotechnology Will not Ensure Food Security, Protect the Environment and Reduce Poverty in the Developing World." *Agbioforum*, vol. 2, nos. 3&4, 1999.

———, and Norman Uphoff. *Report of Bellagio Conference on Sustainable Agriculture*. Cornell International Institute for Food, Agriculture and Development, 1999.

Anderson, Ray. *Mid-Course Correction*. Peregrinzilla Press, Atlanta, Ga., 1998.

Baert, Patrick. *Social Theory in the Twentieth Century*. New York University Press, 1998.

Baltimore, David. "Our genome unveiled." *Nature*, 15 February 2001.

Bardocz, Susan. Panel discussion at conference on "Technology & Globalization." International Forum on Globalization, New York City, February 2001.

Barker, Debi, and Jerry Mander. "Invisible Government." International Forum on Globalization, October 1999.

Barlow, Zenobia, and Margo Crabtree (eds.). *Ecoliteracy: Mapping the Terrain*. Center for Ecoliteracy, Berkeley, Calif., 2000.

Benyus, Janine. *Biomimicry*. Morrow, New York, 1997.

Block, Peter. *Stewardship*. Berrett-Koehler, San Francisco, 1993.

Brown, Lester. *Building a Sustainable Society*. Norton, New York, 1981.

———. "Crossing the Threshold," in *World Watch Magazine*. Worldwatch Institute, Washington, D.C., 1999.

———, et al. *State of the World 2001*. Worldwatch Institute, Washington, D.C., 2001.

Callenbach, Ernest. *Ecology: A Pocket Guide*. University of California Press, Berkeley, 1998.

Capra, Fritjof. *The Tao of Physics.* Shambhala, Boston, 1975; updated fourth edition, 1999.

———. *The Turning Point.* Simon & Schuster, New York, 1982.

———. *Uncommon Wisdom.* Simon & Schuster, New York, 1988.

———. *The Web of Life.* Anchor/Doubleday, New York, 1996.

———. "Is There a Purpose in Nature?" in Anton Markos (ed.), *Is There a Purpose in Nature?* Proceeding of the Prague Workshop, Center for Theoretical Study, Prague, 2000.

———. "Trying to Understand: A Systemic Analysis of International Terrorism." www.fritjofcapra.net, October 2001.

———, and Charlene Spretnak. *Green Politics.* Dutton, New York, 1984.

———, and David Steindl-Rast. *Belonging to the Universe.* Harper, San Francisco, 1991.

———, and Gunter Pauli (eds.). *Steering Business Toward Sustainability.* United Nations University Press, Tokyo, 1995.

Castells, Manuel. *The Information Age*, vol. 1, *The Rise of the Network Society.* Blackwell, 1996.

———. *The Information Age*, vol. 2, *The Power of Identity.* Blackwell, London, 1997.

———. *The Information Age*, vol. 3, *End of Millennium.* Blackwell, London, 1998.

———. "Information Technology and Global Capitalism," in Hutton and Giddens (2000a).

———. "Materials for an Exploratory Theory of the Network Society." *British Journal of Sociology*, vol. 51, no. 1, January/March 2000b.

Chalmers, David J. "Facing Up to the Problem of Consciousness." *Journal of Consciousness Studies*, vol. 2, no. 3, pp. 200–19, 1995.

Chawla, Sarita, and John Renesch (eds.). *Learning Organizations.* Productivity Press, Portland, Ore., 1995.

Churchland, Patricia, and Terrence Sejnowski. *The Computational Brain.* MIT Press, Cambridge, Mass., 1992.

Cohen, Robin, and Shirin Rai. *Global Social Movements.* Athlone Press, 2000.

Crick, Francis. *The Astonishing Hypothesis: The Scientific Search for the Soul.* Scribner, New York, 1994.

Daly, Herman. "Ecological Tax Reform," in Capra and Pauli (1995).

Danner, Mark. "The Lost Olympics." *New York Review of Books*, 2 November 2000.

Davenport, Thomas, and Laurance Prusak. *Working Knowledge.* Harvard Business School Press, 2000.

Dawkins, Richard. *The Selfish Gene.* Oxford University Press, 1976.

De Geus, Arie. *The Living Company.* Harvard Business School Press, 1997a.

———. "The Living Company." *Harvard Business Review*, March–April, 1997b.

Denner, Jason, and Thammy Evans. "Hypercar makes its move." *RMI Solutions.* Rocky Mountain Institute Newsletter, spring 2001.

Dennett, Daniel. *Consciousness Explained.* Little, Brown, New York, 1991.

Depraz, Natalie. "The Phenomenological Reduction as Praxis." *Journal of Consciousness Studies*, vol. 6, no. 2–3, pp. 95–110, 1999.

Dominguez, Joe, and Vicki Robin. *Your Money or Your Life.* Penguin, Harmondsworth, 1999.

Dunn, Seth. "Decarbonizing the Energy Economy," in Brown et al. (2001).

Edelman, Gerald. *The Remembered Present: A Biological Theory of Consciousness.* Basic Books, New York, 1989.

———. *Bright Air, Brilliant Fire.* Basic Books, New York, 1992.

Ehrenfeld, David. "A Techno-Pox Upon the Land." *Harper's Magazine*, October 1997.

Ellul, Jacques. *The Technological Society.* Knopf, New York, 1964.

Faux, Jeff, and Larry Mishel. "Inequality and the Global Economy," in Hutton and Giddens (2000).

Fischer, Aline, Thomas Oberholzer, and Pier Luigi Luisi. "Giant vesicles as models to study the interactions between membranes and proteins." *Biochimica et Biophysica Acta*, vol. 1467, pp. 177–88, 2000.

Fischer, Claude. "Studying Technology and Social Life," in Manuel Castells (ed.). *High Technology, Space, and Society.* Sage, Beverly Hills, Calif., 1985.

Fouts, Roger. *Next of Kin.* William Morrow, New York, 1997.

Galbraith, John Kenneth. *The Anatomy of Power.* Hamish Hamilton, London, 1984.

Gardner, Gary, and Payal Sampat. "Mind over Matter: Recasting the Role of Materials in Our Lives." Worldwatch Paper 144, Worldwatch Institute, Washington, D.C., 1998.

Gelbart, William. "Data bases in Genomic Research." *Science*, 23 October 1998.

Gesteland, Raymond, Thomas Cech, and John Atkins (eds.). *The RNA World.* Cold Spring Harbor Laboratory Press, New York, 1999.

Giddens, Anthony. *Times Higher Education Supplement.* London, 13 December 1996.

Gilbert, Walter. "The RNA World." *Nature*, vol. 319, p. 618, 1986.

Gilmore, David. *Manhood in the Making.* Yale University Press, 1990.

Goldsmith, Edward. "Global Trade and the Environment," in Mander and Goldsmith (1996).

Goodwin, Brian. *How the Leopard Changed Its Spots.* Scribner, New York, 1994.

Gore, Al. *Earth in the Balance.* Houghton Mifflin, New York, 1992.

Gould, Stephen Jay. "Lucy on the Earth in Stasis." *Natural History*, no. 9, 1994.

Halweil, Brian. "Organic Farming Thrives Worldwide," in Lester Brown, Michael Renner, and Brian Halweil (eds.). *Vital Signs 2000.* Norton, New York, 2000.

Havel, Václav. *Disturbing the Peace.* Faber and Faber, London and Boston, 1990.

Hawken, Paul. *The Ecology of Commerce.* HarperCollins, New York, 1993.

———. "N30: WTO Showdown." *Yes!*, spring 2000.

———, Amory Lovins, and Hunter Lovins. *Natural Capitalism.* Little Brown, New York, 1999.

Held, David. *Introduction to Critical Theory.* University of California Press, Berkeley, 1990.

Henderson, Hazel. *Beyond Globalization.* Kumarian Press, West Hartford, Conn., 1999.

Himanen, Pekka. *The Hacker Ethic.* Random House, New York, 2001.

Ho, Mae-Wan. *Genetic Engineering—Dream or Nightmare?* Gateway Books, Bath, U.K., 1998a.

———. "Stop This Science and Think Again." Address to Linnaean Society, London, 17 March 1998b.

Holdrege, Craig. *Genetics and the Manipulation of Life.* Lindisfarne Press, 1996.

Hutton, Will, and Anthony Giddens (eds.). *Global Capitalism.* The New Press, New York, 2000.

Iverson, Jana, and Esther Thelen. "Hand, Mouth and Brain." *Journal of Consciousness Studies,* vol. 6, no. 11–12, pp. 19–40, 1999.

Jackson, Wes. *New Roots for Agriculture.* University of Nebraska Press, 1985.

Johnson, Mark. *The Body in the Mind.* University of Chicago Press, 1987.

Kauffman, Stuart. *At Home in the Universe.* Oxford University Press, 1995.

Keller, Evelyn Fox. *The Century of the Gene.* Harvard University Press, Cambridge, Mass., 2000.

Khor, Martin. "The revolt of developing nations," in "The Seattle Debacle," special issue of *Third World Resurgence.* Penang, Malaysia, December 1999/January 2000.

Kimura, Doreen. "The Neural Basis of Language Qua Gesture," in H. Whitaker and H. A. Whitaker (eds.). *Studies in Linguistics.* vol. 2, Academic Press, San Diego, 1976.

Korten, David. *The Post-Corporate World.* Berrett-Koehler, San Francisco, 1999.

———. *When Corporations Rule the World.* Berrett-Koehler, San Francisco, 1995.

Kranzberg, Melvin, and Carroll Purcell Jr. (eds.). *Technology in Western Civilization.* 2 vols., Oxford University Press, New York, 1967.

Kuttner, Robert. "The Role of Governments in the Global Economy," in Hutton and Giddens (2000).

Lakoff, George. *Women, Fire, and Dangerous Things.* University of Chicago Press, 1987.

———, and Mark Johnson. *Philosophy in the Flesh.* Basic Books, New York, 1999.

Lamb, Marguerite. "Power to the People." *Mother Earth News,* October/November 1999.

Lander, Eric, and Nicholas Schork. "Genetic Dissection of Complex Traits." *Science,* 30 September 1994.

Lappé, Frances Moore, Joseph Collins, and Peter Rosset. *World Hunger: Twelve Myths.* Grove Press, New York, 1998.

Lewontin, Richard. "The Confusion over Cloning." *New York Review of Books,* 23 October 1997.

Losey, J. et al. "Transgenic Pollen Harms Monarch Larvae." *Nature,* 20 May 1999.

Lovelock, James. *Healing Gaia.* Harmony Books, New York, 1991.

Lovins, Amory et al. *Hypercars: Materials, Manufacturing, and Policy Implications.* Rocky Mountain Institute, 1996.

———, and Hunter Lovins. "Frozen Assets?" *RMI Solutions,* Rocky Mountain Institute Newsletter, spring 2001.

Luhmann, Niklas. "The Autopoiesis of Social Systems," in Niklas Luhmann. *Essays on Self-Reference.* Columbia University Press, New York, 1990.

Luisi, Pier Luigi. "About Various Definitions of Life." *Origins of Life and Evolution of the Biosphere,* 28, pp. 613–22, 1998.

———. "Defining the Transition to Life: Self-Replicating Bounded Structures and Chemical Autopoiesis," in W. Stein and F. J. Varela (eds.). *Thinking about Biology.* Addison-Wesley, New York, 1993.

———. "Self-Reproduction of Micelles and Vesicles: Models for the Mechanisms of Life from the Perspective of Compartmented Chemistry," in I. Prigogine and S. A. Rice (eds.). *Advances in Chemical Physics,* vol. xcii. John Wiley, 1996.

Lukes, Steven (ed.). *Power.* New York University Press, 1986.

Mander, Jerry. *In the Absence of the Sacred.* Sierra Club Books, San Francisco, 1991.

Mander, Jerry and Edward Goldsmith (eds.). *The Case Against the Global Economy.* Sierra Club Books, San Francisco, 1996.

Margulis, Lynn. "From Gaia to Microcosm." Lecture at Cortona Summer School, "Science and the Wholeness of Life," August 1998b (unpublished).

———. *Symbiotic Planet.* Basic Books, New York, 1998a.

———, and Dorion Sagan. *Microcosmos.* Published originally in 1986; new edition by University of California Press, Berkeley, 1997.

———, and Dorion Sagan. *What Is Life?* Simon & Schuster, New York, 1995.

Maturana, Humberto. "Biology of Cognition." Published originally in 1970; reprinted in Humberto Maturana and Francisco Varela. *Autopoiesis and Cognition.* D. Reidel, Dordrecht, Holland, 1980.

———. "Biology of Self-Consciousness," in G. Trautteur (ed.). *Consciousness: Distinction and Reflection.* Bibliopolis, Naples, 1995.

Maturana, Humberto. Seminar at members' meeting of the Society for Organizational Learning, Amherst, Mass., June 1998 (unpublished).

———, and Francisco Varela. *The Tree of Knowledge.* Shambhala, Boston, 1987.

McClintock, Barbara. "The Significance of Responses of the Genome to Challenges." 1983 Nobel Lecture, reprinted in Nina Fedoroff and David Botstein (eds.). *The Dynamic Genome.* Cold Spring Harbor Laboratory Press, Cold Spring Harbor, 1992.

McDonough, William, and Michael Braungart. "The Next Industrial Revolution." *Atlantic Monthly*, October 1998.

McGinn, Colin. *The Mysterious Flame.* Basic Books, New York, 1999.

McKibben, Bill. "Some Like it Hot." *New York Review*, 5 July 2001.

McLuhan, Marshall. *Understanding Media.* Macmillan, New York, 1964.

Meadows, Donella. "Scientists Slice Genes as Heedlessly as They Once Split Atoms." *Valley News*, Plainfield, N.H., March 27, 1999.

Medd, William. "Complexity in the Wild: Complexity Science and Social Systems." Ph.D. thesis, Department of Sociology, Lancaster University U.K., March 2000.

Mollison, Bill. *Introduction to Permaculture.* Tagain Publications, Australia, 1991.

Mooney, Patrick. "From Cabbages to Kings," in *Development Dialogue: The Laws of Life.* Dag Hammarskjöld Foundation, Sweden, 1988.

Morgan, Gareth. *Images of Organizations.* Berrett-Koehler, San Francisco, 1998.

Morowitz, Harold. *Beginnings of Cellular Life.* Yale University Press, 1992.

Myers, Norman. *Perverse Subsidies.* International Institute for Sustainable Development, Winnipeg, Manitoba, 1998.

Newman, Peter, and Jeffrey Kenworthy. *Sustainability and Cities.* Island Press, Washington, D.C., 1998.

Nonaka, Ikujiro, and Hirotaka Takeuchi. *The Knowledge-Creating Company.* Oxford University Press, New York, 1995.

Norberg-Hodge, Helena, Todd Merrifield, and Steven Gorelick. "Bringing the Food Economy Home." International Society for Ecology and Culture. Berkeley, California, October 2000.

Núñez, Rafael E. "Eating Soup With Chopsticks: Dogmas, Difficulties and Alternatives in the Study of Conscious Experience." *Journal of Consciousness Studies*, vol. 4, no. 2, pp. 143–66, 1997.

Orr, David. *Ecological Literacy.* State University of New York Press, 1992.

———. *The Nature of Design.* Oxford University Press, New York, 2001.

Pauli, Gunter. "Industrial Clustering and the Second Green Revolution." Lecture at Schumacher College, May 1996 (unpublished).

Pauli, Gunter. *UpSizing.* Greenleaf, 2000.

Penrose, Roger. "The Discrete Charm of Complexity." Keynote Speech at the XXV International Conference of the Pio Manzù Centre, Rimini, Italy, October 1999 (unpublished).

Penrose, Roger. *Shadows of the Mind: A Search for the Missing Science of Consciousness.* Oxford University Press, New York, 1994.

Petzinger, Thomas. *The New Pioneers.* Simon & Schuster, New York, 1999.

Postman, Neil. *Technopoly.* Knopf, New York, 1992.

Pretty, Jules, and Rachel Hine. "Feeding the World with Sustainable Agriculture." U.K. Department for International Development, October 2000.

Proust, Marcel. *In Search of Lost Time*, vol. iv, *Sodom and Gomorrah*. Published originally in 1921; trans. by C. K. Scott Moncrieff and Terence Kilmartin; revised by D. J. Enright. The Modern Library, New York.

Ramonet, Ignacio. "The control of pleasure." *Le Monde Diplomatique*, May 2000.

Register, Richard. *Ecocities.* Berkeley Hills Books, Berkeley (2001).

Register, Richard, and Brady Peeks (eds.). *Village Wisdom/Future Cities.* Ecocity Builders, Oakland, Calif., 1997.

Revonsuo, Antti, and Matti Kamppinen (eds.). *Consciousness in Philosophy and Cognitive Neuroscience.* Lawrence Erlbaum, Hillsdale, N.J., 1994.

Rich, Adrienne. *Of Woman Born.* Norton, New York, 1977.

Robbins, John. *The Food Revolution.* Conari Press, Berkeley, 2001.

Schiller, Dan. "Internet Feeding Frenzy." *Le Monde Diplomatique*, English ed., February 2000.

Searle, John. *Minds, Brains, and Science.* Harvard University Press, Cambridge, Mass., 1984.

———. "The Mystery of Consciousness." *The New York Review of Books*, 2 November and 16 November 1995.

Senge, Peter. *The Fifth Discipline.* Doubleday, New York, 1990.

Senge, Peter. Foreword to Arie de Geus, *The Living Company*, 1996.

Shapiro, James. "Genome System Architecture and Natural Genetic Engineering in Evolution," in Lynn Helena Caporale (ed.). *Molecular Strategies in Biological Evolution, Annals of the New York Academy of Sciences*, vol. 870, 1999.

Shear, Jonathan, and Ron Jevning. "Pure Consciousness: Scientific Eploration of Meditation Techniques." *Journal of Consciousness Studies*, vol. 6, no. 2–3, pp. 189–209, 1999.

Shiva, Vandana. *Biopiracy.* South End Press, Boston, Mass., 1997.

———. "Genetically Engineered Vitamin A Rice: A Blind Approach to Blindness Prevention," in Tokar (2001).

———. "The World on the Edge," in Hutton and Giddens (2000).

Simms, Andrew. "Selling Suicide." Christian Aid Report, May 1999.

Solé, Ricard, and Brian Goodwin. *Signs of Life.* Basic Books, New York, 2000.

Sonea, Sorin, and Maurice Panisset. *A New Bacteriology.* Jones & Bartlett, Sudbury, Mass., 1993.

Soros, George. *The Crisis of Global Capitalism.* Public Affairs, New York, 1998.

Spretnak, Charlene (ed.). *The Politics of Women's Spirituality.* Anchor/Doubleday, New York, 1981.

Stanley, W., S. Ewen, and A. Pusztai. "Effects of Diets Containing Genetically Modified Potatoes . . . on Rat Small Intestines." *Lancet*, 16 October 1999.

Steinbrecher, Ricarda. "What Is Wrong With Nature?" *Resurgence*, May/June 1998.

Steindl-Rast, David. "Spirituality as Common Sense." *The Quest.* Theosophical Society in America, Wheaton, Ill., vol. 3, no. 2, 1990.

Stewart, Ian. *Life's Other Secret.* John Wiley, New York, 1998.

Strohman, Richard. "The Coming Kuhnian Revolution in Biology." *Nature Biotechnology*, vol. 15, March 1997.

Suzuki, David. Panel discussion at conference on "Technology & Globalization." International Forum on Globalization, New York City, February 2001.

———, and Holly Dressel. *From Naked Ape to Superspecies.* Stoddart, Toronto, 1999.

Szostak, Jack, David Bartel, and Pier Luigi Luisi. "Synthesizing Life." *Nature*, vol. 409, nr. 6818, 18 January 2001.

Tokar, Brian (ed.). *Redesigning Life?* Zed, New York, 2001.

Tononi, Giulio, and Gerald Edelman. "Consciousness and Complexity." *Science*, vol. 282, pp. 1846–51, 4 December 1998.

Tuomi, Ilkka. *Corporate Knowledge.* Metaxis, Helsinki, 1999.

Union of International Associations (eds.). *Yearbook of International Organizations*, 4 vols. Saur, Munich, Germany, 2000/2001.

United Nations Development Program (UNDP). *Human Development Report 1996.* Oxford University Press, New York, 1996.

———. *Human Development Report 1999.* Oxford University Press, New York, 1999.

Varela, Francisco. "Neurophenomenology." *Journal of Consciousness Studies*, vol. 3, no. 4, pp. 330–49, 1996a.

———. "Phenomenology in Consciousness Research." Lecture at Dartington Hall, Devon, England, November 1996b (unpublished).

———. "Present-Time Consciousness." *Journal of Consciousness Studies*, vol. 6, no. 2–3, pp. 111–40, 1999.

———. "Resonant Cell Assemblies." *Biological Research*, vol. 28, 81–95, 1995.

———, Evan Thompson, and Eleanor Rosch. *The Embodied Mind.* MIT Press, Cambridge, Mass., 1991.

———, and Jonathan Shear. "First-person Methodologies: What, Why, How?" *Journal of Consciousness Studies*, vol. 6, no. 2–3, pp. 1–14, 1999.

Vermersch, Pierre. "Introspection as Practice." *Journal of Consciousness Studies*, vol. 6, no. 2–3, pp. 17–42, 1999.

Volcker, Paul. "The Sea of Global Finance," in Hutton and Giddens (2000).

Wallace, Alan. "The Buddhist Tradition of Samatha: Methods for Refining and Examining Consciousness." *Journal of Consciousness Studies*, vol. 6, no. 2–3, pp. 175–87, 1999.

Wallach, Lori, and Michelle Sforza. *Whose Trade Organization?* Public Citizen, Washington, D.C., 2001.

Warkentin, Craig, and Karen Mingst. "International Institutions, the State, and Global Civil Society in the Age of the World Wide Web." *Global Governance*, vol. 6, pp. 237–57, 2000.

Watson, James. *The Double Helix*. Atheneum, New York, 1968.

Weatherall, David. "How Much Has Genetics Helped?" *Times Literary Supplement*, London, 30 January 1998.

Wellman, Barry (ed.). *Networks in the Global Village*. Westview Press, Boulder, Colo., 1999.

Wenger, Etienne. *Communities of Practice*. Cambridge University Press, 1998.

———. "Communities of Practice." *Healthcare Forum Journal*, July/August 1996.

Wheatley, Margaret. "The Real Work of Knowledge Management." *Human Resource Information Management Journal*, spring 2001.

———. "Seminar on Self-Organizing Systems." Sundance, Utah, 1997 (unpublished).

———, and Myron Kellner-Rogers. "Bringing Life to Organizational Change." *Journal of Strategic Performance Measurement*, April/May 1998.

Williams, Raymond. *Culture*. Fontana, London, 1981.

Wilson, Don, and Dee Ann Reeder. *Mammal Species of the World*, 2nd ed., Smithsonian Institute Press, 1993.

Windelband, Wilhelm. *A History of Philosophy*. Macmillan, New York, 1901.

Winner, Langdon. *Autonomous Technology*. MIT Press, Cambridge, Mass., 1977.

Winograd, Terry, and Fernando Flores. *Understanding Computers and Cognition*. Addison-Wesley, New York, 1991.

World Commission on Environment and Development. *Our Common Future*. Oxford University Press, New York, 1987.

Zunes, Stephen. "International Terrorism." Institute for Policy Studies, www.fpif.org, September 2001.

| index |